Direct variational methods and
eigenvalue problems in engineering

Monographs and textbooks on mechanics of solids and fluids

editor-in-chief: G. Æ Oravas

Mechanics of elastic stability

editor: H. Leipholz

1. H. LEIPHOLZ
   Theory of elasticity

2. L. LIBRESCU
   Elastostatics and kinetics of anisotropic and heterogeneous shell-type structures

3. C. L. DYM
   Stability theory and its applications to structural mechanics

4. K. HUSEYIN
   Nonlinear theory of elastic stability

5. H. LEIPHOLZ
   Direct variational methods and eigenvalue problems in engineering

Direct variational methods and eigenvalue problems in engineering

H. Leipholz

University of Waterloo
Faculty of Engineering
Solid Mechanics Division
Waterloo, Ontario, Canada

NOORDHOFF INTERNATIONAL PUBLISHING LEYDEN

Softcover reprint of the hardcover 1st edition 1977
A division of A. W. Sijthoff International Publishing Company B.V.,
Leyden, The Netherlands

ISBN 978-90-481-8466-8

Original edition entitled

*'Die direkte Methode der Variationsrechnung und
Eigenwertprobleme der Technik'*

published in 1975 by G. Braun GmbH, Karlsruhe

Contents

Preface

As a consequence of the development of hydro- and aeromechanics, the interest in nonconservative, nonselfadjoint eigenvalue problems has increased remarkably in the recent past. For the solution of problems of this kind, the application of direct methods in the calculus of variations had to be attempted. Still in 1950, A. Pflüger did mention the admissibility of Galerkin's method for nonselfadjoint eigenvalue problems only in form of a conjecture. Challenged by this state of uncertainty in connection with nonclassical eigenvalue problems, the author has devoted himself, like others, to the clarification of related questions. This book is the result of these investigations.

It is well known that in the classical case of conservative problems the methods of Ritz and Galerkin are only variations of each other. Hence, proofs of convergence have been given for the method of Ritz only since such proofs establish at the same time the convergence of Galerkin's method due to the equivalence of the two methods. In the nonclassical case of nonconservative problems it can be shown that there is again equivalence between the methods of Galerkin and Ritz if the method of Ritz is properly modified by including certain boundary terms in the calculation. Moreover, it becomes apparent that the proof of convergence of Galerkin's method can be given easier than for the modified method of Ritz using a corresponding integral equation. This proof of convergence then implies the convergence of the modified method of Ritz by virtue of the equivalence of the two methods.

Special attention has been paid to the task of changing the classical form of Galerkin's method in such a way that this method, like the method of Ritz, functions with admissible functions satisfying only the geometric boundary conditions of the eigenvalue problem. This can again be accomplished by including certain boundary terms in the calculations.

In order to show the larger frame within which the methods of Ritz and Galerkin should be seen, their relationships with other methods, for example the methods of moments, of harmonic balance, of Grammel, and of Trefftz, have been pointed out. For the method of Grammel, it has been proven that this method, like the one of Galerkin, can be applied to nonselfadjoint problems and using coordinate functions which do not satisfy all dynamic boundary conditions.

Variational principles are no more extremum principles in the case of nonconservative, nonselfadjoint problems. In order to show that at least stationary if not extremum functionals may be used in the case of such problems, the method of adjoint systems has been dealt with in Chapter 5. In this chapter, it has been established how a complete analogy to the methods of Ritz and Galerkin of the classical, conservative case can be reached if hybrid equations are being used. Also, a stability theory has been developed by replacing the quadratic functional of the potential energy by a bilinear Hamiltonian functional, which corresponds completely to the classical theory valid for conservative systems, and of which the classical stability theory is a special case.

I am specifically indebted to Dr. Fehrle of Verlag G. Braun, whose friendly support has made the publication of this book possible. Also, I owe my gratitude to Mssrs. R. N. Kapoor, N. Robinson, C. Harris and O. P. Madan who have contributed to the inclusion of instructive examples in the text by carrying out various numerical calculations.

Waterloo, October 1976 *H. Leipholz*

1

Formulation of the basic problem

1.1 The fundamental variational principle

The object of the following considerations are dynamic systems in an (x_i, t) space, where the x_i are cartesian coordinates and t is the time.

It will become apparent that functionals of the kind

$$\mathscr{I}(w_1, \ldots, w_n) = \int_T \int_B F[x_i, t, w_1(x_i, t), \ldots, w_n(x_i, t)]\, dB\, dt \tag{1.1}$$

are essential for the investigation of such systems. In (1.1), the integration is to be carried out over a certain domain B of the x_i subspace and over the interval T of the t axis. The integrand F is a sufficiently often differentiable function of its arguments. The quantities $w_j(x_i, t), j = 1, 2, \ldots, n$, are field functions in the (x_i, t) space.

Assuming that the w_j change by the increments δw_j, to be called *variations* in the following, eq. (1.1) yields

$$\mathscr{I}(w_1 + \delta w_1, \ldots, w_n + \delta w_n)$$
$$= \int_T \int_B F[x_i, t, w_1 + \delta w_1, \ldots, w_n + \delta w_n]\, dB\, dt. \tag{1.2}$$

Applying Taylor's theorem to (1.2) yields

$$\mathscr{I}(w_1 + \delta w_1, \ldots, w_n + \delta w_n) = \mathscr{I}(w_1, \ldots, w_n) + \frac{\partial \mathscr{I}(w_1, \ldots, w_n)}{\partial w_1} \delta w_1$$
$$+ \ldots + \frac{\partial \mathscr{I}(w_1, \ldots, w_n)}{\delta w_n} \delta w_n + \mathscr{R}. \tag{1.3}$$

The remainder $\mathscr{R}$ in (1.3) embraces all terms which involve variations of the second and higher order. It is easy to calculate the increment $\Delta \mathscr{I}$ of the functional which results from the variations δw_j. It reads

$$\Delta \mathscr{I} = \mathscr{I}(w_1 + \delta w_1, \ldots, w_n + \delta w_n) - \mathscr{I}(w_1, \ldots, w_n)$$
$$= \frac{\partial \mathscr{I}}{\partial w_1}\,\delta w_1 + \ldots + \frac{\partial \mathscr{I}}{\partial w_n}\,\delta w_n + \mathscr{R}\,. \tag{1.4}$$

The expression

$$\delta \mathscr{I} = \frac{\partial \mathscr{I}}{\partial w_1}\,\delta w_1 + \ldots + \frac{\partial \mathscr{I}}{\partial w_n}\,\delta w_n\,, \tag{1.5}$$

contained in (1.4), is called the *first variation* of $\mathscr{I}$. Carrying out the operation on the right hand side of (1.5) on (1.1) yields

$$\delta \mathscr{I} = \left[\iint_{T\,B} \frac{\partial F}{\partial w_1}\,\delta w_1 + \ldots + \frac{\partial F}{\partial w_n}\,\delta w_n\right] dB\,dt. \tag{1.6}$$

In the following, we shall deal with technical problems involving bars, plates and shells. Then, the x_i subspace is two-dimensional, i.e., $i = 1, 2$, and moreover, $\mathscr{I} \equiv L$, $F \equiv \mathscr{L}$, where L is *Lagrange's functional* and $\mathscr{L}$ is *Lagrange's density* of the dynamic system concerned. Then, in addition, $n = 7$, and

$$\begin{aligned}
w_1 &= w, & w_5 &\equiv \frac{\partial^2 w}{\partial x_1^2} = w_{x_1 x_1}\,,\\
w_2 &\equiv \frac{\partial w}{\partial x_1} = w_{x_1}\,, & w_6 &\equiv \frac{\partial^2 w}{\partial x_1\,\partial x_2} = w_{x_1 x_2}\,,\\
w_3 &\equiv \frac{\partial w}{\partial x_2} = w_{x_2}\,, & w_7 &\equiv \frac{\partial^2 w}{\partial x_2^2} = w_{x_2 x_2}\,.\\
w_4 &\equiv \frac{\partial w}{\partial t} = w_t\,,
\end{aligned} \tag{1.7}$$

Following a simple transformation,

$$\delta L = \iint_{T\,B} \left[\frac{\partial \mathscr{L}}{\partial w_t}\,\delta w_t + \frac{\partial \mathscr{L}}{\partial w}\,\delta w + \frac{\partial \mathscr{L}}{\partial w_{x_1}}\,\delta w_{x_1} + \ldots + \frac{\partial \mathscr{L}}{\partial w_{x_2 x_2}}\,\delta w_{x_2 x_2}\right] dB\,dt \tag{1.8}$$

is obtained from (1.6).

Now, the term

$$\iint_{T\,B} \frac{\partial \mathscr{L}}{\partial w_t}\, \delta w_t \, dB\, dt,$$

contained in (1.8), may be rewritten by means of integration by parts, which yields for $T = t_1 - t_0$

$$\iint_{T\,B} \frac{\partial \mathscr{L}}{\partial w_t}\, \delta w_t \, dB\, dt = \int_B \left[\frac{\partial \mathscr{L}}{\partial w_t}\, \delta w \right]_{t_0}^{t_1} dB - \iint_{T\,B} \frac{d}{dt}\left(\frac{\partial \mathscr{L}}{\partial w_t} \right) \delta w\, dB\, dt. \quad (1.9)$$

As usual, it is assumed that

$$\delta w\,(t_0) = \delta w\,(t_1) \equiv 0$$

holds true. Then, (1.9) changes into

$$\iint_{T\,B} \frac{\partial \mathscr{L}}{\partial w_t}\, \delta w_t \, dB\, dt = - \iint_{T\,B} \frac{d}{dt}\left(\frac{\partial \mathscr{L}}{\partial w_t} \right) \delta w\, dB\, dt. \quad (1.10)$$

Using (1.10) in (1.8) yields

$$\delta L = \iint_{T\,B} \left[- \frac{d}{dt}\left(\frac{\partial \mathscr{L}}{\partial w_t} \right) \delta w + \left(\frac{\partial \mathscr{L}}{\partial w}\, \delta w + \frac{\partial \mathscr{L}}{\partial w_{x_1}}\, \delta w_{x_1} + \ldots + \frac{\partial \mathscr{L}}{\partial w_{x_2 x_2}}\, \delta w_{x_2 x_2} \right) \right] dB\, dt. \quad (1.11)$$

Introducing the notation

$$\delta \mathscr{L} = \frac{\partial \mathscr{L}}{\partial w}\, \delta w + \frac{\partial \mathscr{L}}{\partial w_{x_1}}\, \delta w_{x_1} + \ldots + \frac{\partial \mathscr{L}}{\partial w_{x_2 x_2}}\, \delta w_{x_2 x_2}, \quad (1.12)$$

equation (1.11) can be simplified to yield

$$\delta L = \iint_{T\,B} \left[- \frac{d}{dt}\left(\frac{\partial \mathscr{L}}{\partial w_t} \right) + \delta \mathscr{L} \right] dB\, dt. \quad (1.13)$$

By definition,

$$\mathscr{L} = \mathscr{T}(w_t) - \mathscr{U}(w, w_{x_1}, \ldots, w_{x_2 x_2}), \quad (1.14)$$

where $\mathcal{T}$ is the *density of the kinetic energy*, which depends on w_t only, and where $\mathcal{U}$ is the *density of the potential energy*, which depends on w and its derivatives with respect to the x_i only. Hence,

$$\frac{\partial \mathcal{L}}{\partial w_t} \equiv \frac{\partial \mathcal{T}}{\partial w_t}, \quad \delta \mathcal{L} \equiv - \delta \mathcal{U}. \tag{1.15}$$

Using (1.15) in (1.13) yields

$$\delta L = - \iint\limits_{T\,B} \left[\frac{d}{dt} \left(\frac{\partial \mathcal{T}}{\partial w_t} \right) \delta w + \delta \mathcal{U} \right] dB\, dt. \tag{1.16}$$

Let now the quantities $P\,[w]$ and

$$Q\,[w] = Q_t\,[w] + Q_x\,[w] \tag{1.17}$$

be introduced. These are quantities having the dimension of a force and which represent differential operators of which P and Q_x operate only with respect to the x_i, while Q_t operates with respect to t only.

The *fundamental variational principle* to be used subsequently throughout the book reads

$$\delta L = - \int\limits_T \left[\int\limits_B Q\,[w]\, \delta w\, dB + \int\limits_R P\,[w]\, \delta w\, dR \right] dt, \tag{1.18}$$

where R denotes the boundary of domain B, and where dR is the line element of the contour of B.

By means of (1.16) and (1.17), the principle can be brought into the form

$$\int\limits_T \left[\int\limits_B \left\{ \frac{d}{dt} \left(\frac{\partial \mathcal{T}}{\partial w_t} \right) \delta w + \delta \mathcal{U} - (Q_t\,[w] + Q_x\,[w])\, \delta w \right\} dB \right.$$
$$\left. - \int\limits_R P\,[w]\, \delta w\, dR \right] dt = 0. \tag{1.19}$$

In the case that $P\,[w] \equiv 0$, $Q\,[w] \equiv 0$, this principle is known as *Hamilton's principle* [1].

1.2 The variational derivative and the derived variational principle

The expression

$$\iint\limits_{T\,B} \delta \mathcal{L}\, dB\, dt = \iint\limits_{T\,B} \left(\frac{\partial \mathcal{L}}{\partial w} \delta w + \frac{\partial \mathcal{L}}{\partial w_{x_1}} \delta w_{x_1} + \frac{\partial \mathcal{L}}{\partial w_{x_2}} \delta w_{x_2} \right.$$

$$+ \frac{\partial \mathscr{L}}{\partial w_{x_1 x_1}} \delta w_{x_1 x_1} + \frac{\partial \mathscr{L}}{\partial w_{x_2 x_2}} \delta w_{x_2 x_2} + \frac{\partial \mathscr{L}}{\partial w_{x_1 x_2}} \delta w_{x_1 x_2} \Bigg) dB\, dt \tag{1.20}$$

shall be rewritten. For this purpose, formula

$$\int_B u \frac{\partial v}{\partial x_i} dB = - \int_B v \frac{\partial u}{\partial x_i} dB + \int_R uv \cos(n, x_i)\, dR \tag{1.21}$$

shall be used which follows from the integral theorem of Gauss. In (1.21), u and v are arbitrary functions, dR has a specific orientation as shown in Figure 1 and therefore possesses an algebraic sign, and in addition,

$$\cos(n, x_1) = \cos\alpha, \ \cos(n, x_2) = \sin\alpha \tag{1.22}$$

is valid, where α is the angle indicated in Figure 1. Setting in (1.21)

$$u = \frac{\partial \mathscr{L}}{\partial w_{x_1}}, \quad v = \delta w,$$

and taking into consideration that

$$\delta w_{x_1} = \delta \frac{\partial w}{\partial x_1} = \frac{\partial}{\partial x_1} \delta w$$

holds true, the relationship

$$\int_T \int_B \frac{\partial \mathscr{L}}{\partial w_{x_1}} \delta w_{x_1}\, dB\, dt$$

$$= - \int_T \int_B \frac{\partial}{\partial x_1} \left(\frac{\partial \mathscr{L}}{\partial w_{x_1}} \right) \delta w\, dB\, dt + \int_T \int_R \frac{\partial \mathscr{L}}{\partial w_{x_1}} \delta w \cos\alpha\, dR\, dt \tag{1.23}$$

is obtained.

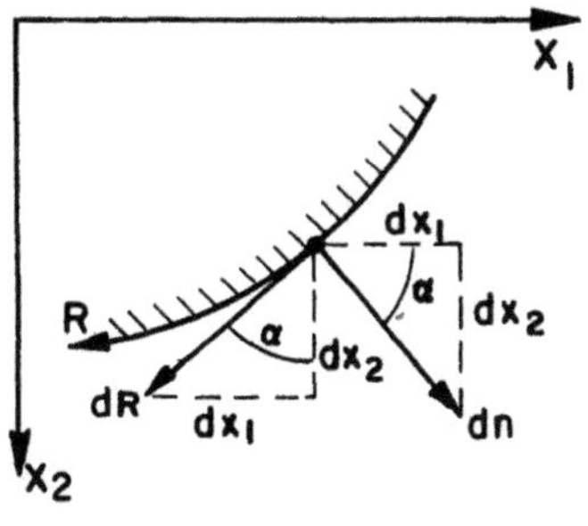

Figure 1

Correspondingly, one arrives at

$$\iint_{T\,B} \frac{\partial \mathcal{L}}{\partial w_{x_2}} \delta w_{x_2} \, dB\,dt$$

$$= - \iint_{T\,B} \frac{\partial}{\partial x_2} \left(\frac{\partial \mathcal{L}}{\partial w_{x_2}} \right) \delta w \, dB\,dt + \iint_{T\,R} \frac{\partial \mathcal{L}}{\partial w_{x_2}} \delta w \sin \alpha \, dR\,dt. \qquad (1.24)$$

Setting in (1.21) on the other hand

$$u = \frac{\partial \mathcal{L}}{\partial w_{x_1}}, \quad v = \delta w_{x_1}$$

and using

$$\delta w_{x_1 x_1} = \frac{\partial}{\partial x_1} \delta w_{x_1},$$

first,

$$\iint_{T\,B} \frac{\partial \mathcal{L}}{\partial w_{x_1 x_1}} \delta w_{x_1 x_1} \, dB\,dt \qquad (1.25)$$

$$= - \iint_{T\,B} \frac{\partial}{\partial x_1} \left(\frac{\partial \mathcal{L}}{\partial w_{x_1 x_1}} \right) \delta w_{x_1} \, dB\,dt + \iint_{T\,R} \frac{\partial \mathcal{L}}{\partial w_{x_1 x_1}} \delta w_{x_1} \cos \alpha \, dR\,dt$$

is obtained.

Now, the first term on the right hand side of (1.25) shall be rewritten by means of (1.21). For that purpose, set

$$u = \frac{\partial}{\partial x_1} \left(\frac{\partial \mathcal{L}}{\partial w_{x_1 x_1}} \right), \quad v = \delta w, \; \delta w_{x_1} = \frac{\partial}{\partial x_1} \delta w,$$

which yields

$$- \iint_{T\,B} \frac{\partial}{\partial x_1} \left(\frac{\partial \mathcal{L}}{\partial w_{x_1 x_1}} \right) \delta w_{x_1} \, dB\,dt \qquad (1.26)$$

$$= \iint_{T\,B} \frac{\partial^2}{\partial x_1^2} \left(\frac{\partial \mathcal{L}}{\partial w_{x_1 x_1}} \right) \delta w \, dB\,dt - \iint_{T\,R} \frac{\partial}{\partial x_1} \left(\frac{\partial \mathcal{L}}{\partial w_{x_1 x_1}} \right) \delta w \cos \alpha \, dR\,dt.$$

Substituting (1.26) back into (1.25) results in

$$\iint_{T\,B} \frac{\partial \mathcal{L}}{\partial w_{x_1 x_1}} \delta w_{x_1 x_1} \, dB\,dt = \iint_{T\,B} \frac{\partial^2}{\partial x_1^2} \left(\frac{\partial \mathcal{L}}{\partial w_{x_1 x_1}} \right) \delta w \, dB\,dt \qquad (1.27)$$

$$- \iint\limits_{T\,R} \frac{\partial}{\partial x_1}\left(\frac{\partial \mathcal{L}}{\partial w_{x_1 x_1}}\right) \delta w \cos\alpha \, dR \, dt + \iint\limits_{T\,R} \frac{\partial \mathcal{L}}{\partial w_{x_1 x_1}} \delta w_{x_1} \cos\alpha \, dR \, dt.$$

However,

$$\delta w_{x_1} = \frac{\partial}{\partial x_1}(\delta w) = -\frac{\partial}{\partial R}(\delta w)\sin\alpha + \frac{\partial}{\partial n}(\delta w)\cos\alpha, \tag{1.28}$$

where $\partial/\partial R$ denotes differentiation along the contour and $\partial/\partial n$ along the normal of boundary R. Hence, for the last term of (1.27),

$$\iint\limits_{T\,R} \frac{\partial \mathcal{L}}{\partial w_{x_1 x_1}} \delta w_{x_1} \cos\alpha \, dR \, dt = - \iint\limits_{T\,R} \frac{\partial \mathcal{L}}{\partial w_{x_1 x_1}} \frac{\partial}{\partial R}(\delta w) \sin\alpha \cos\alpha \, dR \, dt$$

$$+ \iint\limits_{T\,R} \frac{\partial \mathcal{L}}{\partial w_{x_1 x_1}} \frac{\partial}{\partial n}(\delta w) \cos^2\alpha \, dR \, dt \tag{1.29}$$

is obtained. The first term on the right hand side of (1.29) is integrated by parts with respect to R, and the second term is rewritten by means of

$$\frac{\partial}{\partial n}(\delta w) = \delta w_n.$$

These operations lead to

$$\iint\limits_{T\,R} \frac{\partial \mathcal{L}}{\partial w_{x_1 x_1}} \delta w_{x_1} \cos\alpha \, dR \, dt \tag{1.30}$$

$$= \iint\limits_{T\,R} \frac{\partial}{\partial R}\left(\frac{\partial \mathcal{L}}{\partial w_{x_1 x_1}}\right) \delta w \sin\alpha \cos\alpha \, dR \, dt + \iint\limits_{TR} \frac{\partial \mathcal{L}}{\partial w_{x_1 x_1}} \delta w_n \cos^2\alpha \, dR \, dt.$$

Using (1.30) in (1.27) yields finally

$$\iint\limits_{T\,B} \frac{\partial \mathcal{L}}{\partial w_{x_1 x_1}} \delta w_{x_1 x_1} \, dB \, dt = \iint\limits_{T\,B} \frac{\partial^2}{\partial_{x_1^2}}\left(\frac{\partial \mathcal{L}}{\partial w_{x_1 x_1}}\right) \delta w \, dB \, dt$$

$$- \iint\limits_{T\,R} \frac{\partial}{\partial x_1}\left(\frac{\partial \mathcal{L}}{\partial w_{x_1 x_1}}\right) \delta w \cos\alpha \, dR \, dt + \iint\limits_{T\,R} \frac{\partial}{\partial R}\left(\frac{\partial \mathcal{L}}{\partial w_{x_1 x_1}}\right)$$

$$\delta w \sin\alpha \cos\alpha \, dR \, dt + \iint\limits_{TR} \frac{\partial \mathcal{L}}{\partial w_{x_1 x_1}} \delta w_n \cos^2\alpha \, dR \, dt. \tag{1.31}$$

Correspondingly, using (1.21), (1.22) and

$$\delta w_{x_2} = \frac{\partial}{\partial x_2}(\delta w) = \frac{\partial}{\partial R}(\delta w)\cos\alpha + \frac{\partial}{\partial n}(\delta w)\sin\alpha, \tag{1.32}$$

relationship

$$\iint\limits_{T\,B} \frac{\partial \mathscr{L}}{\partial w_{x_2 x_2}} \delta w_{x_2 x_2}\, dB\, dt = \iint\limits_{T\,B} \frac{\partial^2}{\partial x_2^2}\left(\frac{\partial \mathscr{L}}{\partial w_{x_2 x_2}}\right)\delta w\, dB\, dt$$
$$- \iint\limits_{T\,R} \frac{\partial}{\partial x_2}\left(\frac{\partial \mathscr{L}}{\partial w_{x_2 x_2}}\right)\delta w \sin\alpha\, dR\, dt \tag{1.33}$$
$$- \iint\limits_{T\,R} \frac{\partial}{\partial R}\left(\frac{\partial \mathscr{L}}{\partial w_{x_2 x_2}}\right)\delta w \sin\alpha\cos\alpha\, dR\, dt + \iint\limits_{T\,R} \frac{\partial \mathscr{L}}{\partial w_{x_2 x_2}}\delta w_n \sin^2\alpha\, dR\, dt$$

is obtained.

An extensive calculation using (1.21), (1.22), (1.28), and (1.32) yields in the same way

$$\iint\limits_{T\,B} \frac{\partial \mathscr{L}}{\partial w_{x_1 x_2}} \delta w_{x_1 x_2}\, dB\, dt = \iint\limits_{T\,B} \frac{\partial^2}{\partial x_1 \partial x_2}\left(\frac{\partial \mathscr{L}}{\partial w_{x_1 x_2}}\right)\delta w\, dB\, dt$$
$$- \frac{1}{2}\iint\limits_{T\,R} \frac{\partial}{\partial x_2}\left(\frac{\partial \mathscr{L}}{\partial w_{x_1 x_2}}\right)\delta w \cos\alpha\, dR\, dt$$
$$- \frac{1}{2}\iint\limits_{T\,R} \frac{\partial}{\partial x_1}\left(\frac{\partial \mathscr{L}}{\partial w_{x_1 x_2}}\right)\delta w \sin\alpha\, dR\, dt$$
$$+ \frac{1}{2}\iint\limits_{T\,R} \frac{\partial}{\partial R}\left(\frac{\partial \mathscr{L}}{\partial w_{x_1 x_2}}\right)\delta w \sin^2\alpha\, dR\, dt - \frac{1}{2}\iint\limits_{T\,R} \frac{\partial}{\partial R}\left(\frac{\partial \mathscr{L}}{\partial w_{x_1 x_2}}\right)\delta w \cos^2\alpha$$
$$dR\, dt + \iint\limits_{TR} \frac{\partial \mathscr{L}}{\partial w_{x_1 x_2}}\delta w_n \sin\alpha\cos\alpha\, dR\, dt. \tag{1.34}$$

Introducing the so-called *variational derivative* of L

$$\frac{\delta L}{\delta w} = \frac{\partial \mathscr{L}}{\partial w} - \frac{\partial}{\partial x_1}\left(\frac{\partial \mathscr{L}}{\partial w_{x_1}}\right) - \frac{\partial}{\partial x_2}\left(\frac{\partial \mathscr{L}}{\partial w_{x_2}}\right) + \frac{\partial^2}{\partial x_1^2}\left(\frac{\partial \mathscr{L}}{\partial w_{x_1 x_1}}\right)$$
$$+ \frac{\partial^2}{\partial x_2^2}\left(\frac{\partial \mathscr{L}}{\partial w_{x_2 x_2}}\right) + \frac{\partial^2}{\partial x_1 \partial x_2}\left(\frac{\partial \mathscr{L}}{\partial w_{x_1 x_2}}\right), \tag{1.35}$$

and the expression

$$V(w, \delta w) = \left\{\frac{\partial \mathscr{L}}{\partial w_{x_1}}\cos\alpha + \frac{\partial \mathscr{L}}{\partial w_{x_2}}\sin\alpha - \frac{\partial}{\partial x_1}\left(\frac{\partial \mathscr{L}}{\partial w_{x_1 x_1}}\right)\cos\alpha\right.$$

$$- \frac{\partial}{\partial x_2}\left(\frac{\partial \mathscr{L}}{\partial w_{x_2 x_2}}\right)\sin\alpha - \frac{1}{2}\left[\frac{\partial}{\partial x_1}\left(\frac{\partial \mathscr{L}}{\partial w_{x_1 x_2}}\right)\sin\alpha\right.$$

$$\left. + \frac{\partial}{\partial x_2}\left(\frac{\partial \mathscr{L}}{\partial w_{x_1 x_2}}\right)\cos\alpha\right] + \frac{\partial}{\partial R}\left[\left(\frac{\partial \mathscr{L}}{\partial w_{x_1 x_1}} - \frac{\partial \mathscr{L}}{\partial w_{x_2 x_2}}\right)\sin\alpha\cos\alpha\right.$$

$$\left.\left. + \frac{1}{2}\frac{\partial \mathscr{L}}{\partial w_{x_1 x_2}}\sin^2\alpha - \frac{1}{2}\frac{\partial \mathscr{L}}{\partial w_{x_1 x_2}}\cos^2\alpha\right]\right\}\delta w$$

$$+ \left\{\frac{\partial \mathscr{L}}{\partial w_{x_1 x_1}}\cos^2\alpha + \frac{\partial \mathscr{L}}{\partial w_{x_2 x_2}}\sin^2\alpha + \frac{\partial \mathscr{L}}{\partial w_{x_1 x_2}}\sin\alpha\cos\alpha\right\}\delta w_n, \quad (1.36)$$

the integral (1.20) can be brought into the form

$$\int_T\int_R \delta \mathscr{L}\, dB\, dt = \int_T\left[\int_B \frac{\delta L}{\delta w}\,\delta w\, dB + \int_R V(w, \delta w)\, dR\right] dt \quad (1.37)$$

by means of (1.23), (1.24), (1.31), (1.33), (1.34), (1.35), and (1.36). Moreover, introducing the kinetic energy

$$T = \int_B \mathscr{T} dB, \quad (1.38)$$

the potential energy

$$U = \int_B \mathscr{U} dB \quad (1.39)$$

of the dynamic system, using

$$L = \int_B \mathscr{L} dB, \quad (1.40)$$

and taking (1.14) into account, the relationship

$$L = T - U = \int_B \mathscr{T}(w_t)\, dB - \int_B \mathscr{U}(w, w_{x_1}, \ldots, w_{x_2 x_2})\, dB \quad (1.41)$$

results. But then, according to (1.35) and (1.41),

$$\frac{\delta L}{\delta w} \equiv -\frac{\delta U}{\delta w} = -\frac{\partial \mathscr{U}}{\partial w} + \frac{\partial}{\partial x_1}\left(\frac{\partial \mathscr{U}}{\partial w_{x_1}}\right) + \frac{\partial}{\partial x_2}\left(\frac{\partial \mathscr{U}}{\partial w_{x_2}}\right) - \frac{\partial^2}{\partial x_1^2}\left(\frac{\partial \mathscr{U}}{\partial w_{x_1 x_1}}\right)$$

$$- \frac{\partial^2}{\partial x_2^2}\left(\frac{\partial \mathscr{U}}{\partial w_{x_2 x_2}}\right) - \frac{\partial^2}{\partial x_1 \partial x_2}\left(\frac{\partial \mathscr{U}}{\partial w_{x_1 x_2}}\right). \quad (1.42)$$

Applying (1.14) to (1.36) yields also

$$V(w, \delta w) \equiv -W(w, \delta w) = -\left\{\frac{\partial \mathscr{U}}{\partial w_{x_1}} \cos\alpha + \frac{\partial \mathscr{U}}{\partial w_{x_2}} \sin\alpha - \frac{\partial}{\partial x_1}\left(\frac{\partial \mathscr{U}}{\partial w_{x_1x_1}}\right)\cos\alpha - \frac{\partial}{\partial x_2}\left(\frac{\partial \mathscr{U}}{\partial w_{x_2x_2}}\right)\sin\alpha - \frac{1}{2}\left[\frac{\partial}{\partial x_1}\left(\frac{\partial \mathscr{U}}{\partial w_{x_1x_2}}\right)\sin\alpha + \frac{\partial}{\partial x_2}\frac{\partial \mathscr{U}}{\partial w_{x_1x_2}}\right]\cos\alpha + \frac{\partial}{\partial R}\left[\left(\frac{\partial \mathscr{U}}{\partial w_{x_1x_1}} - \frac{\partial \mathscr{U}}{\partial w_{x_2x_2}}\right)\sin\alpha\cos\alpha + \frac{1}{2}\frac{\partial \mathscr{U}}{\partial w_{x_1x_2}}\sin^2\alpha - \frac{1}{2}\frac{\partial \mathscr{U}}{\partial w_{x_1x_2}}\cos^2\alpha\right]\right\}\delta w - \left\{\frac{\partial \mathscr{U}}{\partial w_{x_1x_1}}\cos^2\alpha + \frac{\partial \mathscr{U}}{\partial w_{x_2x_2}}\sin^2\alpha + \frac{\partial \mathscr{U}}{\partial w_{x_1x_2}}\sin\alpha\cos\alpha\right\}\delta w_n. \tag{1.43}$$

Inserting (1.42) and (1.43) in (1.37) yields

$$\iint_{T\,B} \delta\mathscr{L}\, dB\, dt = -\int_T\left[\int_B \frac{\delta U}{\delta w}\,\delta w\, dB + \int_R W(w, \delta w)\, dR\right] dt. \tag{1.44}$$

Taking (1.15) into account,

$$\iint_{T\,B} \delta\mathscr{U}\, dB\, dt = \int_T\left[\int_B \frac{\delta U}{\delta w}\,\delta w\, dB + \int_R W(w, \delta w)\, dR\right] dt \tag{1.45}$$

is obtained in place of (1.44). It may be pointed out that $\delta\mathscr{U}$ is to be determined in the same way as $\delta\mathscr{L}$ follows from (1.12).

By means of (1.45), the fundamental variational principle can be written to yield

$$\int_T\left[\int_B\left\{\frac{d}{dt}\left(\frac{\partial\mathscr{T}}{\partial w_t}\right) + \frac{\delta U}{\delta w} - (Q_t[w] + Q_x[w])\right\}\delta w\, dB + \int_R\left\{W(w, \delta w) - P[w]\,\delta w\right\} dR\right] dt = 0. \tag{1.46}$$

In this form it shall be called the *derived variational principle*, since it has been derived from the fundamental principle by using the integral formula (1.21).

Such a procedure would correspond to *integration by parts* in an one-dimensional space.

If the principle (1.46) is applied such that $W(w, \delta w) - P[w]\,\delta w = 0$ on R, which is equivalent to satisfying certain boundary conditions, then,

$$\iint_{T\,B} \left\{ \frac{d}{dt}\left(\frac{\partial \mathscr{T}}{\partial w_t}\right) + \frac{\delta U}{\delta w} - (Q_t[w] + Q_x[w]) \right\} \delta w \, dB \, dt = 0 \tag{1.47}$$

remains from (1.46). This relationship shall be called the *principle of Galerkin.*

1.3 The boundary value problem equivalent to the variational principles

The dynamic systems to be considered are either supported along their boundaries in a prescribed way or they are free along their boundaries. The supported part of the boundary shall subsequently be donated by R_L and the free part by R_F. The support can either consist in a free support along the part R_{LS} of R_L or the system can be clamped along the part R_{LE} of R_L.

As to be shown in examples, *geometric* or *essential boundary conditions* are imposed on the dynamic system along part R_L of its boundary. These are

$$\begin{aligned} w &= w^0 \text{ on } R_{LS}, \\ w &= w^0 \text{ and } w_n = w_n^0 \text{ on } R_{LE}. \end{aligned} \tag{1.48}$$

In (1.48), w^0 and w_n^0 are prescribed functions which may degenerate into constants which also can have the value zero. In general it is assumed that, like w, the varied functions $\overline{w} = w + \delta w$ satisfy the boundary conditions (1.48). Therefore, at least,

$$\begin{aligned} \delta w &= 0 \text{ on } R_{LS}, \\ \delta w &= 0 \text{ and } \delta w_n = 0 \text{ on } R_{LE}. \end{aligned} \tag{1.49}$$

The complete set of boundary conditions is always of such a nature that in addition to (1.49),

$$W(w, \delta w) - P[w]\,\delta w \equiv 0 \text{ on } R. \tag{1.50}$$

There is no condition for w_n on R_{LS}, hence, δw_n is arbitrary on that part of the boundary. There are no conditions set for w and w_n on R_F. Therefore, δw and δw_n are arbitrary on R_F. Considering the arbitrariness of these quantities on R_{LS} and R_F, using the definition of $W(w, \delta w)$ as given by (1.43), and attempting to satisfy (1.50), the following *dynamic* or *natural boundary conditions* are obtained:

$$\frac{\partial \mathscr{U}}{\partial w_{x_1 x_1}} \cos^2\alpha + \frac{\partial \mathscr{U}}{\partial w_{x_2 x_2}} \sin^2\alpha + \frac{\partial \mathscr{U}}{\partial w_{x_1 x_2}} \sin\alpha\cos\alpha = 0 \text{ on } R_{LS} \text{ and } R_F,$$

$$\frac{\partial \mathscr{U}}{\partial w_{x_1}} \cos\alpha + \frac{\partial \mathscr{U}}{\partial w_{x_2}} \sin\alpha - \frac{\partial}{\partial x_1}\left(\frac{\partial \mathscr{U}}{\partial w_{x_1 x_1}}\right)\cos\alpha - \frac{\partial}{\partial x_2}\left(\frac{\partial \mathscr{U}}{\partial w_{x_2 x_2}}\right)\sin\alpha$$

$$-\frac{1}{2}\left[\frac{\partial}{\partial x_1}\left(\frac{\partial \mathscr{U}}{\partial w_{x_1 x_2}}\right)\sin\alpha + \frac{\partial}{\partial x_2}\left(\frac{\partial \mathscr{U}}{\partial w_{x_1 x_2}}\right)\cos\alpha\right]$$

$$+\frac{\partial}{\partial R}\left[\left(\frac{\partial \mathscr{U}}{\partial w_{x_1 x_1}} - \frac{\partial \mathscr{U}}{\partial w_{x_2 x_2}}\right)\sin\alpha\cos\alpha + \frac{1}{2}\frac{\partial \mathscr{U}}{\partial w_{x_1 x_2}} \sin^2\alpha\right.$$

$$\left.-\frac{1}{2}\frac{\partial \mathscr{U}}{\partial w_{x_1 x_2}}\cos^2\alpha\right] - P[w] = 0 \text{ on } R_F. \qquad (1.51)$$

After these introductory remarks, it may be assumed that the fundamental principle (1.19) has been applied to determine the function w in such a manner that this principle be satisfied.

Firstly, the search for w may have been carried out in the space of functions that are sufficiently continuous and differentiable and do satisfy *all* the boundary conditions (1.48) and (1.51) so that (1.50) holds true on R. The function w thus defined may then be used in (1.19). Moreover, by means of (1.21), the principle (1.19) may be transformed into the derived variational principle. This has to be obtained in its special form (1.47), since it has been assumed that *all* boundary conditions are satisfied. Therefore (1.50) holds. Since the variation δw is arbitrary in the domain B, equation (1.47) can be correct only if w satisfies the *differential equation*

$$\frac{d}{dt}\left(\frac{\partial \mathscr{T}}{\partial w_t}\right) + \frac{\delta U}{\delta w} - (Q_t[w] + Q_x[w]) = 0. \qquad (1.52)$$

It may be recalled that w satisfies also the *boundary conditions* (1.48) and (1.51) by assumption.

On the basis of these statements, the true meaning of the variational principle (1.19) becomes apparent: it serves for determining the solution w and the eigenvalues* of the equivalent boundary value problem (1.48), (1.51) and (1.52). It may be mentioned in addition, that (1.52) is called the *Euler equation* of (1.19). This equation represents the *equation of motion* of the system concerned, speaking in terms of mechanics. Obviously, the practical importance of variational principles consists in the fact that they can be used to evaluate the solution w and the eigenvalues of (1.52) in a convenient way without requiring that the equation of motion itself be solved. As will be shown, solution and eigenvalues are found by applying the so called *direct method of the calculus of variations.*

Secondly, the fundamental variational principle may be used to determine w by searching in the enlarged space of the sufficiently continuous and differentiable functions which do satisfy *the geometric boundary conditions (1.48) only.* The thus defined w may then be used, in (1.19), as before, and this principle transformed. Obviously, the result is this time

$$\int_T \left[\int_B \left\{ \frac{d}{dt}\left(\frac{\partial \mathcal{T}}{\partial w_t}\right) + \frac{\delta U}{\delta w} - (Q_t\,[w] + Q_x\,[w]) \right\} \delta w\, dB \right.$$

$$\left. + \int_{R_{LS}} S\,[w]\,\delta w_n\, dR + \int_{R_F} \left\{ W(w, \delta w) - P\,[w]\,\delta w \right\} dR \right] dt = 0,$$

$$S\,[w] = \frac{\partial\, \mathcal{U}}{\partial w_{x_1 x_1}} \cos^2\alpha + \frac{\partial\, \mathcal{U}}{\partial w_{x_2 x_2}} \sin^2\alpha + \frac{\partial\, \mathcal{U}}{\partial w_{x_1 x_2}} \sin\alpha \cos\alpha. \qquad (1.53)$$

But since δw is arbitrary in B and δw_n on R_{LS}, δw, δw_n on R_F are likewise arbitrary, equation (1.53) can be correct only if again the Euler equation (1.52) and if in addition the dynamic boundary conditions (1.51) are satisfied. The latter statement follows from the fact that the integrands in the line integrals over R_{LS} and R_F in (1.53) must vanish due to arbitrariness of δw and δw_n. Finally, it may be recalled that the geometric boundary conditions are already satisfied by definition. Therefore, it follows again that the boundary value problem (1.48), (1.51) and (1.52) is equivalent to the fundamental variational principle (1.19). Moreover, it has become evident that *it is sufficient to apply the*

* The eigenvalues mentioned here will be considered in detail only later, in Section 1.5 and in following sections. For the sake of clarity it may however be mentioned already now, that the boundary value problems to be tackled will be *homogeneous* ones. Therefore, a solution w will exist for them only, if certain parameters involved in the problem assume specific values, the so-called eigenvalues. Hence, eigenvalues and the corresponding *eigensolutions* w, with which we are dealing presently, are closely related.

fundamental variational principle (1.19) to functions which do satisfy the geometric boundary conditions only. The principle works in such a manner, that it imposes the fulfillment of the dynamic boundary conditions automatically on the solution w. This is a fact of great practical importance, as an application of the principle is greatly facilitated if it can be applied to functions in a function space which is as large as possible.

Corresponding deliberations are now carried out for the derived variational principle (1.46). Firstly, it may again be assumed that the function w had been determined by means of this principle under observation of *all* boundary conditions. For such a w, (1.50) holds true. Consequently, (1.47) is immediately satisfied instead of (1.46). But since δw is arbitrary in B, (1.47) can only be valid, if w is a solution of differential equation (1.52). The result of these deliberations is therefore the following: if the variational principle (1.46) or the equivalent principle (1.47) of Galerkin is applied to functions in a space whose elements are sufficiently continuous and differentiable and which satisfy *all* boundary conditions (1.48), (1.51) on R, then a function w is obtained, which is a solution of the equivalent boundary value problem (1.48), (1.51) and (1.52).

If the derived variational principle is to be applied using the advantage of a wider function space whose elements satisfy *the geometric boundary conditions only*, then, (1.46) has obviously to be used in the form

$$\int_T \left[\int_B \left\{ \frac{d}{dt}\left(\frac{\partial \mathcal{T}}{\partial w_t}\right) + \frac{\delta U}{\delta w} - (Q_t[w] + Q_x[w]) \right\} \delta w \, dB \int_{R_{LS}} S[w]\,\delta w_n \, dR \right.$$

$$\left. + \int_{R_F} \left\{ W(w, \delta w) - P[w]\,\delta w \right\} dR \right] dt = 0,$$

$$S[w] = \frac{\partial \mathcal{U}}{\partial w_{x_1 x_1}} \cos^2\alpha + \frac{\partial \mathcal{U}}{\partial w_{x_2 x_2}} \sin^2\alpha + \frac{\partial \mathcal{U}}{\partial w_{x_1 x_1}} \sin\alpha\cos\alpha, \qquad (1.54)$$

since one part of the line integral vanishes in (1.46) due to the assumption that w satisfies a priori the geometric boundary conditions. Under certain circumstances, $R_{LS} \equiv 0$ may hold true in (1.54). Then, an additional line integral vanishes in that relationship. Having determined a function w for which (1.54) becomes an identity, that is only possible if w is a solution of (1.52) and satisfies in addition boundary conditions (1.51) on R_{LS} and R_F. The first assertion must hold true because δw is arbitrary in B. The second assertion must be correct because δw and δw_n on R_F and δw_n on R_{LS} are all arbitrary. Again, the dynamic boundary conditions have been enforced through applying the derived principle in its form (1.54). As a whole, also in this case, the principle (1.54) yields a function w, which is a solution to the equivalent boundary value problem (1.48), (1.51), (1.52).

Summarizing, the following can be stated: when working in a space of functions which satisfy *all* boundary conditions, the fundamental variational principle (1.19) is equivalent to Galerkin's principle (1.47). On the other hand, when working in a space of functions which satisfy the *geometric boundary conditions only*, the fundamental variational principle (1.19) corresponds to the derived variational principle in its form (1.54), and the dynamic boundary conditions are then imposed on function w. In all possible cases, function w obtained through the variational principles is the solution of the equivalent boundary value problem (1.48), (1.51), (1.52).

The very aim of all efforts is to determine the function w and the eigenvalues of the boundary value problem, since these quantities describe in a characteristic manner the motion of the mechanical system concerned. This aim can be reached in two ways when using the variational principles. Firstly, one may evaluate function w and the eigenvalues by using the so called *direct method of the calculus of variations*, which will be considered extensively in the following. Secondly, one may use the variational principles indirectly, i.e., for the purpose of deriving the equivalent boundary value problem. This problem will then be solved in the classic way for w and the eigenvalues. Such an approach is called the *indirect method of the calculus of variations*, since the quantities of interest are being determined not by means of the principles themselves but by the boundary value problem which is equivalent to them.

1.4 Equations of motion and boundary conditions

As shown in the preceding section, the equivalent boundary value problem (1.48), (1.51), (1.52) is obtained by applying (1.21) to the variational principle (1.19). The differential equation involved is, as already mentioned, known as Euler's equation in terms of mathematics. From the point of view of.mechanics, this differential equation represents the *equation of motion* of the system concerned.

In the following, beam, plate, and shell problems shall be considered. For the first two sets of problems,

$$\mathcal{T} = \frac{1}{2}\,\mu w_t^2\,, \tag{1.55}$$

where μ is the constant line mass density for beam problems, and the constant mass density per surface unit for plate and shell problems. As far as nonconservative forces are concerned, the following shall hold: $P[w]$ and $Q_x[w]$ are supposed to be *follower forces*, i.e., forces which change their lines of action according to the deformation of the system to which they are applied. Therefore,

$$P[w] = -d_j w_{x_j}, Q_x[w] = -k_j w_{x_j}, \tag{1.56}$$

where d_j, k_j are components of load vectors which have to be determined from case to case. The quantity

$$Q_t[w] = -\beta w_t \tag{1.57}$$

is supposed to have the meaning of a friction, if taken into consideration at all. In general, this shall not be the case, specifically not in this section. According to (1.52) and (1.42), the density $\mathscr{U}$ of the potential energy will be especially relevant for the derivation of the equation of motion. Therefore, $\mathscr{U}$ has to be specified very carefully for the respective system. This quantity will also determine in the first place the dynamic boundary conditions according to (1.51).

As a first example, consider the rod shown in Figure 2. Its length be l. Its constant flexural rigidity be α and its constant line mass density be μ. The rod is assumed to carry the conservative uniformly distributed *unidirectional* load g, which bears that designation because it retains its line of action regardless of the deformation of the rod. The density of the potential energy reads

$$\mathscr{U} = \frac{1}{2}[\alpha w_{xx}^2 - g(l-x) w_x^2]. \tag{1.58}$$

From (1.55) follows

$$\frac{d}{dt}\frac{\partial \mathscr{T}}{\partial w_t} = \mu w_{tt}, \tag{1.59}$$

and from (1.42), (1.58) follows

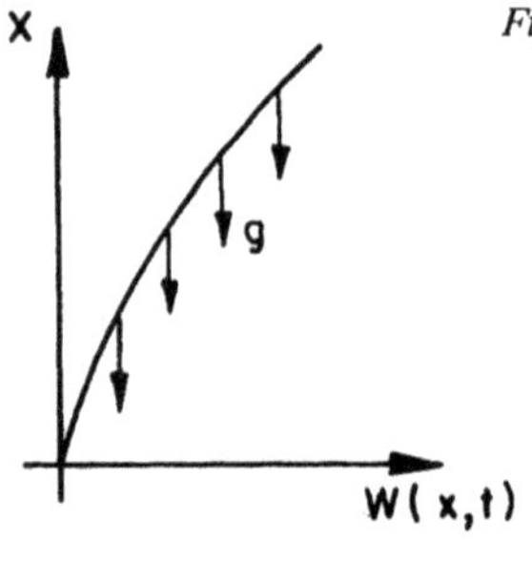

Figure 2

Figure 3

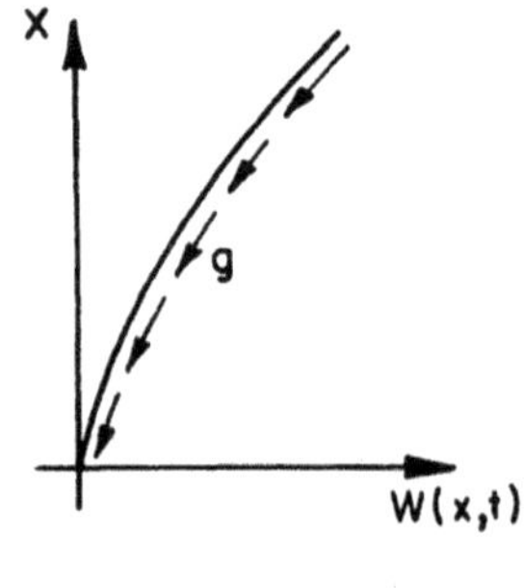

$$\frac{\delta U}{\delta w} = [g(l-x)\, w_x]_x + w_{xxxx}, \tag{1.60}$$

Q_t [w] shall be disregarded. Since the external load is supposed to be unidirectional and therefore conservative, $P[w] = 0$, $Q_x[w] = 0$ must hold true. With all this, (1.52) yields

$$\mu w_{tt} + \alpha w_{xxxx} + g\,[(l-x)\, w_x]_x = 0 \tag{1.61}$$

as the equation of motion of the rod. The rod is clamped at the lower end. Therefore,

$$w = 0, \;\; w_x = 0 \text{ at } x = 0 \tag{1.62}$$

according to (1.48). The rod is free at the upper end and satisfies there the dynamic boundary conditions (1.51). By virtue of $P[w] = 0$, $\cos\alpha = 1$, $\sin\alpha = 0$, and (1.58), these boundary conditions are

$$w_{xxx} = w_{xx} = 0 \text{ at } x = l. \tag{1.63}$$

The second example is the rod in Figure 3, which is subjected to the nonconservative, uniformly distributed follower load g. Everything else is supposed to be as in the previous example, except that

$$Q_x[w] = -g w_x \tag{1.64}$$

holds true. By means of (1.59), (1.60), (1.64), and $P[w] \equiv 0$, $Q_t[w] \equiv 0$, (1.52) yields the equation of motion

$$\mu w_{tt} + \alpha w_{xxxx} + g(l-x)\, w_{xx} = 0 \tag{1.65}$$

of the nonconservative rod problem shown in Figure 3. The boundary conditions (1.62) and (1.63) remain valid in this case too.

Next, the plate of constant thickness h displayed in Figure 4 shall be considered. It is assumed that all its edges are freely supported and that it is subjected to uniformly distributed, unidirectional compressive loads N along the edges $x_1 = 0$ and $x_1 = a$.

The density of the kinetic energy is given by (1.55). The density of the potential energy reads

$$\mathcal{U} = \frac{D}{2}\left[(w_{x_1x_1} + w_{x_2x_2})^2 + 2(1-\nu)(w_{x_1x_2}^2 - w_{x_1x_1} w_{x_2x_2}) - \frac{N}{D}\, w_{x_1}^2\right]. \tag{1.66}$$

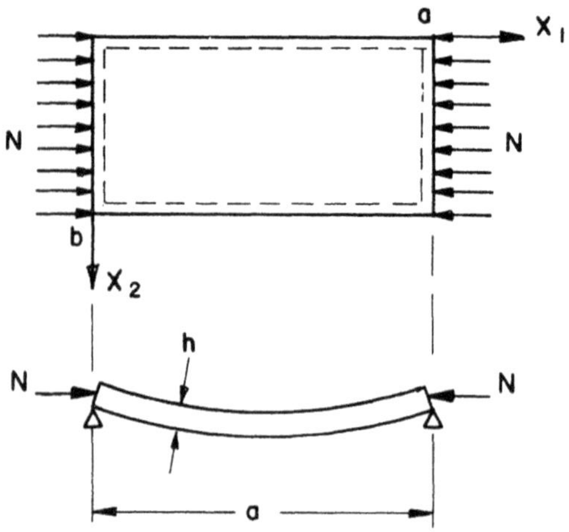

Figure 4

In (1.66), D is the bending stiffness of the plate, ν Poisson's ratio, and $w(x_1, x_2)$ the deflection of the middle plane of the plate.

From (1.55) follows (1.59), and from (1.42), (1.66) follows

$$\frac{\delta U}{\delta w} = D\,[w_{x_1x_1x_1x_1} + 2w_{x_2x_2x_1x_1} + w_{x_2x_2x_2x_2}] + Nw_{x_1x_1} =$$
$$= D\Delta^2 w + Nw_{x_1x_1}\,. \tag{1.67}$$

$Q_t\,[w]$ is to be neglected. Since the external load was assumed to be unidirectional and is therefore conservative, also $Q_x\,[w]$ and $P\,[w]$ do not appear. Considering all this,

$$\mu w_{tt} + D\Delta^2 w + Nw_{x_1x_1} = 0 \tag{1.68}$$

is derived from (1.52) as the equation of motion of the plate in Figure 4.

The boundary conditions read as follows: the plate is supported along all edges. Therefore,

$$w = 0 \text{ along } x_1 = 0, x_1 = a, x_2 = 0, x_2 = b \tag{1.69}$$

according to (1.48). That are the geometric boundary conditions. The plate is freely supported along all edges. Therefore, (1.51) must be taken into consideration for the case R_{LS}.

Along $x_1 = 0, x_1 = a$, $|\cos\alpha| = 1$, $\sin\alpha = 0$. Thus, the first dynamic boundary condition reads

$$\frac{\partial\,\mathscr{U}}{\partial w_{x_1x_1}} = 0,$$

which is

$$w_{x_1x_1} + \nu w_{x_2x_2} = 0 \text{ along } x_1 = 0, x_1 = a \tag{1.70}$$

according to (1.66).

Along $x_2 = 0, x_2 = b$, $|\sin\alpha| = 1$, $\cos\alpha = 0$, which yields

$$\frac{\partial \mathscr{U}}{\partial w_{x_2x_2}} = 0,$$

and by means of (1.66), the second dynamic boundary condition

$$w_{x_2x_2} + \nu w_{x_1x_1} = 0 \text{ along } x_2 = 0, x_2 = b. \tag{1.71}$$

The plate shown in Figure 5 is subjected to a nonconservative edge load along $x_1 = a$. All other edges are freely supported except this one, which is free. Therefore, it can be assumed that the load N along this edge follows the deformation and is always tangential to the middle plane of the plate. This loading case has been treated by 0. Petterson [2]. Assuming that angle θ_a is small in Figure 5, relationship (1.66) remains valid, and (1.52) yields again the differential equation (1.68), since $Q_t\,[w] = 0$ is supposed to hold true, and since $Q_x\,[w] = 0$. This latter condition is correct because the nonconservative load is an *edge load* and is not distributed over the plate. Due to the edge load,

$$P\,[w] = -N w_{x_1} \text{ along } x_1 = a. \tag{1.72}$$

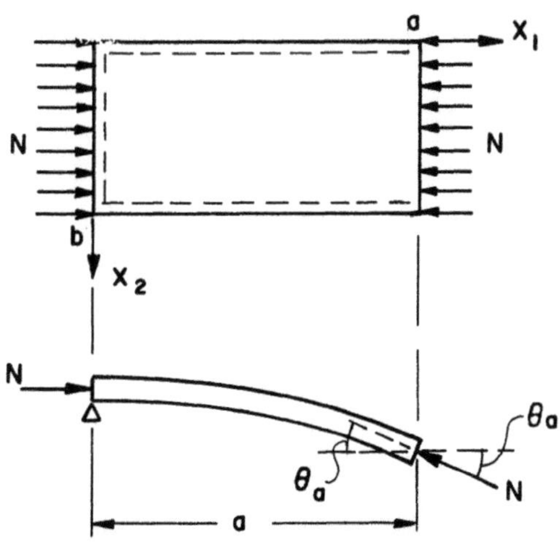

Figure 5

Therefore, and because of a different kind of support, the boundary conditions differ from those in the preceding example.

On R_{LS},

$$w = 0 \text{ along } x_1 = 0, x_2 = 0, x_2 = b, \tag{1.73}$$

which are the geometric boundary conditions. The dynamic boundary conditions follow from (1.51), (1.66).

Along $x_1 = 0$, R_{LS} has to be taken into account, and $\cos\alpha = -1$, $\sin\alpha = 0$. Hence,

$$\frac{\partial \mathscr{U}}{\partial w_{x_1x_1}} = 0$$

follows from (1.51), which by virtue of (1.66) yields

$$w_{x_1x_1} + \nu w_{x_2x_2} = 0 \text{ along } x_1 = 0. \tag{1.74}$$

Along $x_2 = 0$, $x_2 = b$, again, R_{LS} is valid, and $|\sin\alpha| = 1$, $\cos\alpha = 0$. Therefore,

$$\frac{\partial \mathscr{U}}{\partial w_{x_2x_2}} = 0$$

is obtained from (1.51), which yields

$$w_{x_2x_2} + \nu w_{x_1x_1} = 0 \text{ along } x_2 = 0, x_2 = b. \tag{1.75}$$

Along $x_1 = a$, one has R_F as well as $P[w] = -Nw_{x_1}$, $\cos\alpha = 1$, $\sin\alpha = 0$, $(\partial/\partial R) = (\partial/\partial x_2)$. Using all this in (1.51), and taking (1.66) into account, yields

$$\frac{\partial \mathscr{U}}{\partial w_{x_1x_1}} = 0,\; w_{x_1x_1} + \nu w_{x_2x_2} = 0 \text{ along } x_1 = a, \tag{1.76}$$

$$\frac{\partial \mathscr{U}}{\partial w_{x_1}} - \frac{\partial}{\partial x_1}\left(\frac{\partial \mathscr{U}}{\partial w_{x_1x_1}}\right) - \frac{\partial}{\partial x_2}\left(\frac{\partial \mathscr{U}}{\partial w_{x_1x_2}}\right) - P[w] = 0,$$

$$w_{x_1x_1x_1} + (2-\nu)\, w_{x_1x_2x_2} = 0 \text{ along } x_1 = a, \tag{1.77}$$

respectively. The last boundary condition, which expresses the fact that the edge $x_1 = a$ is free of forces, is at the same time an indication of the nonconservative character of the problem.

Also the plate in Figure 6 is loaded nonconservatively. In this case, a tangential follower load g, which is parallel to x_1, is uniformly distributed over the

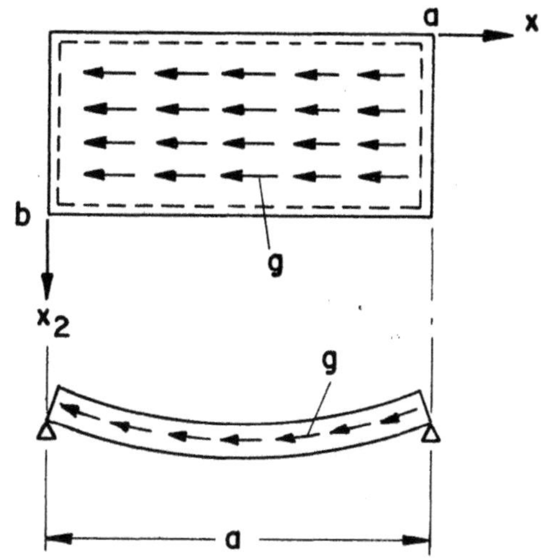

Figure 6

middle surface of the plate. Consequently,

$$\mathscr{U} = \frac{D}{2}\left[(w_{x_1x_1} + w_{x_2x_2})^2 + 2(1-\nu)(w_{x_1x_2}^{\;2} - w_{x_1x_1} w_{x_2x_2}) - \frac{g}{D}\,(a-x_1) w_{x_1}^2 \right] \tag{1.78}$$

and

$$Q_x\,[w] = -g w_{x_1}, P\,[w] = 0. \tag{1.79}$$

$Q_t\,[w] = 0$ is supposed to hold. According to (1.42),

$$\frac{\delta U}{\delta w} = D\,[w_{x_1x_1x_1x_1} + 2w_{x_1x_1x_2x_2} + w_{x_2x_2x_2x_2}] + g\,[(a-x_1)w_{x_1}]_{x_1}. \tag{1.80}$$

Using (1.80), (1.79), (1.59), and the fact that Q_t vanishes, in (1.52) yields the equation of motion

$$\mu w_{tt} + D\Delta^2\, w + g\,(a-x_1)\, w_{x_1x_1} = 0, \tag{1.81}$$

while the boundary conditions (1.69), (1.70), (1.71) remain valid.

As a conclusion, it may be shown how the equations of motion of the circular cylindrical shell shown in Figure 7 can be derived. The shell is subjected to an axial, unidirectional compressive load $N = \sigma h$, uniformly distributed along the edges of a cut perpendicular to the shell axis. Let a be the radius of the middle surface of the shell, h the constant thickness of the shell, μ the mass per surface unit, and u, v, w the displacements of the points of the middle surface in direction of the generator and of the tangent and normal at the circumference.

Moreover, the strains of the middle surface are

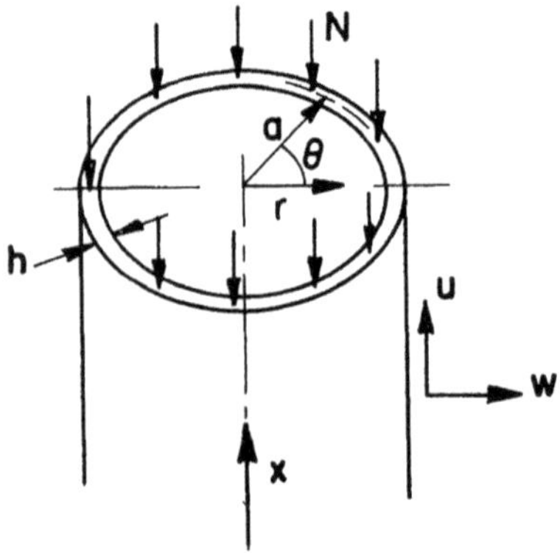

Figure 7

$$
\begin{aligned}
e_x &= \frac{\partial u}{\partial x} = u_x, \\
e_\theta &= \frac{1}{a}\left(\frac{\partial v}{\partial \theta} + w\right) = \frac{v_\theta + w}{a}, \\
e_{x\theta} &= \frac{\partial v}{\partial x} + \frac{1}{a}\,\frac{\partial u}{\partial \theta} = v_x + \frac{u_\theta}{a}.
\end{aligned}
\tag{1.82}
$$

The rates of curvature of the middle surface are

$$
\begin{aligned}
\kappa_x &= \frac{\partial^2 w}{\partial x^2} = w_{xx}, \\
\kappa_\theta &= \frac{1}{a}\,\frac{\partial}{\partial \theta}\left(\frac{v}{a} + \frac{1}{a}\,\frac{\partial w}{\partial \theta}\right) = \frac{v_\theta}{a^2} + \frac{w_{\theta\theta}}{a^2}, \\
\kappa_{x\theta} &= \frac{1}{a}\left(\frac{\partial v}{\partial x} + 2\,\frac{\partial^2 w}{\partial x \partial \theta}\right) = \frac{v_x}{a} + \frac{2}{a}\, w_{x\theta}.
\end{aligned}
\tag{1.83}
$$

Hence, the density of the potential energy of the internal and external forces

$$
\begin{aligned}
\mathscr{U} = {} & \frac{Eha}{2\,(1-\nu^2)}\left[e_x^2 + e_\theta^2 + 2\nu e_x e_\theta + \frac{1}{2}(1-\nu)\, e_{x\theta}^2\right] \\
& + \frac{Eh^3 a}{24\,(1-\nu^2)}\left[\kappa_x^2 + \kappa_\theta^2 + 2\nu\kappa_x\kappa_\theta + \frac{1}{2}\,(1-\nu)\,\kappa_{x\theta}^2\right] - \frac{1}{2}\, Na\,(u_x^2 + v_x^2 + w_x^2)
\end{aligned}
$$

changes into

$$\mathscr{U} = \frac{Eha}{2(1-\nu^2)}\left[u_x^2 + \left(\frac{v_\theta + w^2}{a}\right) + 2\nu u_x\left(\frac{v_\theta + w}{a}\right) + \frac{1}{2}(1-\nu)\left(v_x\right.\right.$$

$$\left.\left. + \frac{u_\theta}{a}\right)^2\right] + \frac{Eh^3 a}{24(1-\nu^2)}\left[w_{xx}^2 + \left(\frac{v_\theta}{a^2} + \frac{w_{\theta\theta}}{a^2}\right)^2 + 2\nu w_{xx}\left(\frac{v_\theta + w_{\theta\theta}}{a^2}\right)\right.$$

$$\left. + \frac{1}{2}(1-\nu)\left(\frac{v_x + 2w_{x\theta}}{a}\right)^2\right] - \frac{1}{2}Na\,(u_x^2 + v_x^2 + w_x^2), \tag{1.84}$$

when using (1.82), (1.83) in $\mathscr{U}$.

The density of the kinetic energy is

$$\mathscr{T} = \frac{\mu_a}{2}(u_t^2 + v_t^2 + w_t^2). \tag{1.85}$$

Since the load is conservative,

$$Q_t[u] = Q_t[v] = Q_t[w] = 0, Q_x[u] = Q_x[v] = Q_x[w] = 0,$$

and since the densities $\mathscr{U}$ and $\mathscr{T}$ do now depend on the three functions u, v, and w, the following three equations of motion are obtained from (1.52), respectively,

$$\begin{aligned}
&\frac{d}{dt}\left(\frac{\partial\mathscr{T}}{\partial u_t}\right) + \frac{\delta U}{\delta u} = 0,\\
&\frac{d}{dt}\left(\frac{\partial\mathscr{T}}{\partial v_t}\right) + \frac{\delta U}{\partial v} = 0,\\
&\frac{d}{dt}\left(\frac{\partial\mathscr{T}}{\partial w_t}\right) + \frac{\delta U}{\delta x} = 0.
\end{aligned} \tag{1.86}$$

Using (1.42), and taking (1.84), (1.85), (1.86) into account, yields

$$\begin{aligned}
&\mu a u_{tt} - \frac{\partial}{\partial x}\frac{\partial\mathscr{U}}{\partial u_x} - \frac{\partial}{\partial\theta}\frac{\partial\mathscr{U}}{\partial u_\theta} = 0,\\
&\mu a v_{tt} - \frac{\partial}{\partial x}\frac{\partial\mathscr{U}}{\partial v_x} - \frac{\partial}{\partial\theta}\frac{\partial\mathscr{U}}{\partial v_\theta} = 0,
\end{aligned} \tag{1.87}$$

$$\mu a w_{tt} + \frac{\partial \mathscr{U}}{\partial w} - \frac{\partial}{\partial x}\frac{\partial \mathscr{U}}{\partial w_x} + \frac{\partial^2}{\partial x^2}\frac{\partial \mathscr{U}}{\partial w_{xx}} + \frac{\partial^2}{\partial \theta^2}\frac{\partial \mathscr{U}}{\partial w_{\theta\theta}}$$
$$+ \frac{\partial^2}{\partial x \partial \theta}\frac{\partial \mathscr{U}}{\partial w_{x\theta}} = 0.$$

Obviously,

$$\frac{\partial \mathscr{U}}{\partial u_x} = \frac{Eha}{2\,(1-\nu^2)}\left[2u_x + 2\nu\left(\frac{v_\theta + w}{a}\right)\right] - Nau_x,$$

$$\frac{\partial \mathscr{U}}{\partial u_\theta} = \frac{Eha}{2\,(1-\nu^2)}\left[(1-\nu)\left(v_x + \frac{u_\theta}{a}\right)\frac{1}{a}\right].$$

Using these relationships in the first line of (1.87) yields after simple rearrangement

$$u_{xx} + \frac{1-\nu}{2a^2}u_{\theta\theta} + \frac{1+\nu}{2a}v_{x\theta} + \frac{\nu}{a}w_x - \frac{Nu_{xx}(1-\nu^2)}{Eh} = \frac{\mu(1-\nu^2)}{Eh}u_{tt}. \tag{1.88}$$

That is the first equation of motion.

Furthermore,

$$\frac{\partial \mathscr{U}}{\partial v_x} = \frac{Eha}{2(1-\nu^2)}(1-\nu)\left(v_x + \frac{u_\theta}{a}\right) + \frac{Eh^3 a}{24(1-\nu^2)}\left[(1-\nu)\left(\frac{v_x + 2w_{x\theta}}{a}\right)\frac{1}{a}\right]$$
$$- Nav_x,$$

$$\frac{\partial \mathscr{U}}{\partial v_\theta} = \frac{Eha}{2(1-\nu^2)}\left[2\left(\frac{v_\theta + w}{a}\right)\frac{1}{a} + \frac{2\nu}{a}u_x\right] + \frac{Eh^3 a}{24(1-\nu^2)}\left[2\left(\frac{v_\theta + w_{\theta\theta}}{a^2}\right)\frac{1}{a^2}\right.$$
$$\left. + \frac{2\nu}{a^2}w_{xx}\right].$$

Using these relationships in (1.87), the second line of this set of equations yields after a simple rearrangement

$$\frac{1-\nu}{2}v_{xx} + \frac{1+\nu}{2}u_{x\theta} + \frac{v_{\theta\theta}}{a^2} + \frac{w_\theta}{a^2} + \frac{h^2}{12}\left[\frac{1-\nu}{2}\frac{v_{xx}}{a^2} + \frac{w_{xx\theta}}{a^2} + \frac{v_{\theta\theta}}{a^4} + \frac{w_{\theta\theta\theta}}{a^4}\right]$$
$$- \frac{Nv_{xx}\,(1-\nu^2)}{Eh} = \frac{\mu\,(1-\nu^2)}{Eh}v_{tt}. \tag{1.89}$$

That is the second equation of motion.

Finally, one finds that

$$\frac{\partial \mathscr{U}}{\partial w} = \frac{Eha}{2(1-\nu^2)}\left[\frac{2}{a}\left(\frac{v_\theta + w}{a}\right) + \frac{2\nu ux}{a}\right],$$

$$\frac{\partial \mathscr{U}}{\partial w_x} = -Naw_x,$$

$$\frac{\partial \mathscr{U}}{\partial w_{xx}} = \frac{Eh^3 a}{24(1-\nu^2)}\left[2w_{xx} + 2\nu\left(\frac{v_\theta + w_{\theta\theta}}{a^2}\right)\right],$$

$$\frac{\partial \mathscr{U}}{\partial w_{\theta\theta}} = \frac{Eh^3 a}{24(1-\nu^2)}\left[2\left(\frac{v_\theta + w_{\theta\theta}}{a^2}\right)\frac{1}{a^2} + \frac{2\nu}{a^2} w_{xx}\right],$$

$$\frac{\partial \mathscr{U}}{\partial w_{x\theta}} = \frac{Eh^3 a}{24(1-\nu^2)}\left[(1-\nu)\left(\frac{v_x + 2w_{x\theta}}{a}\right)\frac{2}{a}\right].$$

Using these relationships in the third line of (1.87) yields after rearrangement

$$\frac{v_\theta}{a^2} + \frac{w}{a^2} + \frac{\nu}{a}u_x + \frac{h^2}{12}\left[w_{xxxx} + \frac{2w_{xx\theta\theta}}{a^2} + \frac{w_{\theta\theta\theta\theta}}{a^4} + \frac{v_{xx\theta}}{a^2} + \frac{v_{\theta\theta\theta}}{a^4}\right]$$

$$+ \frac{Nw_{xx}(1-\nu^2)}{Eh} = -\frac{\mu(1-\nu^2)}{Eh}w_{tt}. \tag{1.90}$$

That is the third equation of motion.

For the complete description of the problem, the triple of equations, (1.88), (1.89), (1.90), has to be supplemented by a set of boundary conditions. These shall not be specified here in full generality since there are essential simplifications possible when the theory is applied to practical cases.

For example, let the cylindrical shell be treated, whose two edges are freely supported. Moreover, assume that the shell buckles in a rotational symmetric form if the load is rotational symmetric. Due to symmetry, $v \equiv 0$, and all derivatives with respect to θ vanish. Therefore, equation (1.90) can be reduced to

$$\frac{w}{a^2} + \frac{\nu}{a}u_x + \frac{h^2}{12}w_{xxxx} + \frac{Nw_{xx}(1-\nu^2)}{Eh} = -\frac{\mu(1-\nu^2)}{Eh}w_{tt}. \tag{1.91}$$

It is plausible to set $e_x = -\nu e_\theta$. According to (1.82), using also $v_\theta = 0$, this assumption leads to

$$u_x = -\frac{\nu w}{a}. \tag{1.92}$$

Using (1.92) in (1.91) yields after some calculation.

$$w_{xxxx} + \kappa^2 w_{xx} + \alpha^2 w = -\frac{\mu(1-\nu^2)}{Eh} w_{tt}, \tag{1.93}$$

where

$$\kappa^2 = \frac{12(1-\nu^2)N}{Eh^3} = \frac{12(1-\nu^2)\sigma}{Eh^2}, \alpha^2 = \frac{12(1-\nu^2)}{a^2 h^2}. \tag{1.94}$$

The boundary conditions are

$$\begin{aligned} w &= 0 \text{ for } x = 0, x = L, \\ w_{xx} &= 0 \text{ for } x = 0, x = L, \end{aligned} \tag{1.95}$$

where L is the length of the cylindrical shell.

Although in the following only the structure of differential equation (1.93) will be of interest, it may be worthwhile to remark that the buckling stress σ_k of the shell can be determined for $w_{tt} = 0$ from

$$\kappa^2 = 2\alpha. \tag{1.96}$$

Using (1.94) in (1.96) yields

$$\sigma_k = \frac{Eh}{a} \frac{1}{\sqrt{3(1-\nu^2)}}. \tag{1.97}$$

That is the result already given by A. and L. Föppl [3].

1.5 Eigenvalue problems. Direct methods

In all cases in Section 1.4, a certain type of problem has arisen, which shall now be specified. For this purpose, it shall be referred to equations (1.61), (1.62), (1.63) and (1.65), (1.62), (1.63) for beams, equations (1.68), (1.69), (1.70), (1.71) and (1.68), (1.73), (1.74), (1.75), (1.76), (1.77) and (1.81), (1.69), (1.70), (1.71) for plates, equations (1.93), (1.95) for shells. In all these cases one has to deal with a homogeneous differential equation of the type

$$(T_t w) + (M_x w) = 0 \tag{1.98}$$

and with homogeneous boundary conditions of the kind

$$(G_x w)_R = 0. \tag{1.99}$$

Here, T_t is a linear differential operator acting only on the variable t, and M_x as well as G_x are linear differential operators with respect to the variables x_i, $i = 1,2$. In order to keep the following considerations simple, the restriction to a single independent variable x may be introduced. Then, only beam and arch like structural elements are covered, yet these are of great practical importance. How the investigations could be extended to include two-dimensional structures like plates and shells has been pointed out by S. L. Sobolev [4].

It is obvious that problems of the kind of (1.98), (1.99) admit of particular solutions of the form

$$w(x,t) = f(t)\,y(x). \tag{1.100}$$

By means of (1.100), equations (1.98), (1.99) yield

$$\frac{Tf}{f} = -\frac{My}{y} = -\lambda \tag{1.101}$$

$$f(Gy)_R = 0, \tag{1.102}$$

where λ is a constant yet to be determined. By means of (1.101), (1.102), firstly,

$$My - \lambda y = 0, \tag{1.103}$$

$$(Gy)_R = 0 \tag{1.104}$$

is being obtained. That is an *eigenvalue-boundary value problem*, whose solution yields the *eigenvalue*

$$\lambda = \lambda(g). \tag{1.105}$$

In (1.105) it has been assumed that λ is a function of a structural parameter g contained in the operator M.

Secondly, the differential equation

$$Tf + \lambda f = 0 \tag{1.106}$$

follows from (1.101). In (1.106), λ can be considered to be a given quantity. Since T represents the linear operator

$$T = c\,\frac{d^2}{dt^2}\,, \tag{1.107}$$

where c is a constant, the solution of (1.106) is

$$f(t) = e^{i\omega t}. \tag{1.108}$$

Using this solution in (1.106), and taking (1.107) into account, yields

$$-c\omega^2 + \lambda = 0,$$

which is the 'frequency equation' of the problem. Hence,

$$\omega = \sqrt{\lambda/c}. \tag{1.109}$$

The time behaviour of ω is determined by the factor f. According to (1.108) this behaviour is a stable one if ω is a real quantity. This is ensured if λ is a positive and real quantity as follows from (1.109).

The basic problem to be dealt with subsequently can be sketched as follows: Firstly, the eigenvalue-boundary value problem (1.103), (1.104) must be solved. This task will be affected by the structure of the operator M. In the classical literature, e.g., [5], it is assumed that M is of the order $2m$, positive definite and selfadjoint. Here it shall be allowed that the operator M of the order $2m$ be nonselfadjoint and even nonlinear (quasilinear).

In a broader sense, the basic problem may also include stability considerations of the system. However, we shall restrict ourselves to the immediate problem, which is the calculation of the eigenvalue. The question of stability shall be touched occasionally only.

For the calculation of the solution w and of the eigenvalues, the already mentioned direct method of the calculus of variations will be used. This method consists in using the trial solution

$$w(x,t) = f(t) \sum_i a_i \phi_i(x) \tag{1.110}$$

in the fundamental variational principle (1.19) or in the derived variational principal (1.46), thus avoiding the use of the equivalent boundary value problem (1.48), (1.51), (1.52). The direct method yields a so called *characteristic equation* for the eigenvalue λ and, by means of the eigenvalue, the coefficients a_i of the solution (1.110) except *one*, which remains undetermined. As long as λ is real, the coefficients a_i will be real. But then the function

$$y(x) = \sum_i a_i \phi_i(x) \tag{1.111}$$

will also be real, since one can assume that for the expansion of $y(x)$ into a series, a system of real, *orthonormal coordinate functions* $\phi_i(x)$ has been used. Hence,

$$\int_R \phi_i(x)\, \phi_k(x)\, dx = \delta_{ik}, \tag{1.112}$$

where δ_{ik} is the *Kronecker-symbol*.

In the domain of real functions $y(x)$ which satisfy the boundary conditions (1.104), an expansion of the kind (1.111) is indeed justified. This can be shown in the following way:

$$H\phi - \lambda\phi = 0, \tag{1.113}$$

$$(G\phi)_R = 0 \tag{1.114}$$

be introduced in which the differential operator H is supposed to be selfadjoint, positive definite and of the same order $2m$ as the operator M in (1.103). Also, it is assumed that the boundary conditions correspond to those in (1.104). As known, the problem (1.113), (1.114) yields a countable, infinite set of eigenfunctions ϕ_i, which can be used as coordinate functions in (1.111). This fact holds true, because $y(x)$, as long as it is real, can be used as a *comparison function* of (1.113), (1.114), since it satisfies due to the equivalence of (1.114) to (1.104), all boundary conditions of the auxiliary problem. For comparison functions, the *expansion theorem* in the formulation of L. Collatz [6] holds true. According to this theorem, the series (1.111) and the series obtained by differentiating (1.111) ν times ($\nu = 0,1, \ldots, m-1$) converge uniformly and absolutely inside of the boundary R.

Obviously, the direct method, which uses the trial solution (1.111), is essentially based on the assumption that a real function $y(x)$ exists, which can be expanded into the series (1.111). However, that is not a restriction of generality, since the systems are practically of interest only below the so called 'flutter limit'. This limit is reached when λ, and therefore y, do reach the limit of their domain. Beyond the flutter limit, the mechanical system will suffer large deformations which have to be investigated in a special way, especially by means of a nonlinear approach. But then, the linear differential equation (1.98), and the equations (1.103), (1.106) following from (1.98), are invalid.

2

The method of Ritz

2.1 The classical case. Conservative problems

Consider the fundamental variational principle (1.19) under the assumption of vanishing nonconservative forces Q_t [w], Q_x [w], and P [w]. Then, the principle changes into

$$\int_T \left[\int_B \left\{ \frac{d}{dt}\left(\frac{\partial \mathcal{T}}{\partial w_t}\right) \delta w + \delta \, \mathcal{U} \right\} dB \right] dt = 0 \tag{2.1}$$

which is Hamilton's principle as already mentioned in Section 1.1. According to considerations in Section 1.5,

$$\frac{d}{dt}\left(\frac{\partial \mathcal{T}}{\partial w_t}\right) = Tw. \tag{2.2}$$

Using this relationship in (2.1) yields

$$\int_T \left[\int_B \{(Tw) \, \delta w + \delta \, \mathcal{U} \} dB \right] dt = 0. \tag{2.3}$$

Restricting oneself to one-dimensional problems, i.e., to problems involving a single independent variable x only,

$$\mathcal{U} = \frac{1}{2} \sum_{k=0}^{2} h_k (x) \, [w^{(k)}(x, t)]^2 \,. \tag{2.4}$$

In (2.4), h_k (x) are variable (or possibly constant) coefficients, and

$$w^{(k)} (x, t) = \frac{\partial^k w}{\partial x^k} \,, \; k = 0, 1, 2.$$

By definition, one is dealing with problems having solutions of the form

$$w(x,t) = f(t)\,y(x). \tag{1.100}$$

Substituting (1.100) and (2.4) back in (2.3) yields

$$\int_T f^2\,dt \int_B \left\{\frac{Tf}{f}\,y\delta y + \delta\left(\frac{1}{2}\sum_{k=0}^{2} h_k(x)\,[y^{(k)}(x)]^2\right)\right\} dB = 0. \tag{2.5}$$

Using (1.101) in (2.5) results in

$$\int_T f^2\,dt \int_B \left\{-\lambda y\delta y + \delta\left(\frac{1}{2}\sum_{k=0}^{2} h_k(x)\,[y^{(k)}(x)]^2\right)\right\} dB = 0. \tag{2.6}$$

This relationship can hold true only if

$$\int_B \left\{-\lambda y\delta y + \delta\left(\frac{1}{2}\sum_{k=0}^{2} h_k(x)\,[y^{(k)}(x)]^2\right)\right\} dB = 0. \tag{2.7}$$

This condition can be rewritten to yield

$$\delta\left[\int_B \frac{1}{2}\left\{-\lambda y^2 + \sum_{k=0}^{2} h_k(x)\,[y^{(k)}(x)]^2\right\} dB\right] = 0, \tag{2.8}$$

which is equivalent to

$$\int_B \frac{1}{2}\left\{-\lambda y^2 + \sum_{k=0}^{2} h_k(x)\,[y^{(k)}(x)]^2\right\} dB = \text{stationary}. \tag{2.9}$$

Hence, the fundamental variational principle is a true 'extremum principle' in the classical case.

The method of Ritz consists in its classical form in substituting for $y(x)$ the series expansion

$$y(x) = \sum_i a_i\,\phi_i(x) \tag{1.111}$$

in (2.8). It is assumed that the coordinate functions $\phi_i(x)$ do satisfy at least the geometrical boundary conditions. This is sufficient for enforcing automatically the dynamic boundary conditions when applying the fundamental variational principal, as shown in Section 1.3.

For the practical application of the method,

$$\int_B \left\{-\lambda y\delta y + \sum_{k=0}^{2} h_k\,y^{(k)}\,\delta y^{(k)}\right\} dB = 0 \tag{2.10}$$

is derived from (2.7) or (2.8). Then,

$$y^{(k)} = \sum_i a_i \phi_i^{(k)}, \quad \delta y^{(k)} = \frac{\partial y^{(k)}}{\partial a_i} \delta a_i = \phi_i^{(k)} \delta a_i \tag{2.11}$$

is used in (2.10) according to (1.111). The result is

$$\int_B \left\{ -\lambda (\sum_i a_i \phi_i)(\sum_j \phi_j \delta a_j) + \sum_{k=0}^{2} h_k (\sum_i a_i \phi_i^{(k)})(\sum_j \phi_j^{(k)} \delta a_j) \right\} dB = 0. \tag{2.12}$$

Because of the arbitrariness of the variations δa_j, the system of equations

$$\int_B \left\{ -\lambda (\sum_i a_i \phi_i) \phi_j + \sum_{k=0}^{2} h_k (\sum_i a_i \phi_i^{(k)}) \phi_j^{(k)} \right\} dB = 0, \; j = 1,2,3, \ldots, \tag{2.13}$$

follows from (2.12). Due to (1.112), and the additivity of the integration, equations (2.13) can be changed into

$$\sum_i a_i \left\{ -\lambda \delta_{ij} + \int_B \sum_{k=0}^{2} h_k \phi_i^{(k)} \phi_j^{(k)} \, dB \right\} = 0, \quad j = 1,2,3, \ldots . \tag{2.14}$$

Usually,

$$h_1(x) = g h_1^*(x), \tag{2.15}$$

where g is a 'load parameter'. Considering this fact, then

$$\int_B \sum_{k=0}^{2} h_k \phi_i^{(k)} \phi_j^{(k)} dB = a_{ij}(g), \tag{2.16}$$

and (2.14) changes into

$$\sum_i a_i \{ -\lambda \delta_{ij} + a_{ij}(g) \} = 0, \qquad j = 1,2,3, \ldots . \tag{2.17}$$

The set of equations (2.17) is equivalent to the linear, algebraic system

$$\sum_i a_i \rho_{ij} = 0, \qquad j = 1,2,3, \ldots , \tag{2.18}$$

where

$$\rho_{ij} = a_{ij}(g) - \lambda \delta_{ij} \tag{2.19}$$

has been used as an abbreviation. The system (2.18) yields nontrivial solutions for the coefficients a_i of (1.111) only if

$$\det \rho_{ij} = 0. \tag{2.20}$$

Since the ρ_{ij} are functions of g and λ, according to (2.19), one is lead finally to the eigenvalue equation

$$F(g,\lambda) = 0 \tag{2.21}$$

by applying the method of Ritz according to (2.20). This equation shows that indeed

$$\lambda = \lambda(g). \tag{1.105}$$

The eigenvalue λ is therefore a function of the parameter g, which, through its influence on λ, affects the stability behaviour of the system, as indicated in Section 1.5.

Plotting the curve defined by (2.21) in a (g, λ) – plane, one arrives at two basic, possible kinds of traces when restricting oneself to the first two branches of the curve: In Figure 8a, the two branches are monotonically increasing with always negative slope so that they intersect the positive g axis at the points $g_{1\,\text{crit}}$, $g_{2\,\text{crit}}$, penetrating afterwards the domain of *negative* λ – values. As already mentioned, stability is lost when λ ceases to be positive (and real). Therefore, the values $g_{1\,\text{crit}}$, $g_{2\,\text{crit}}$ define the stability limit. Since λ is the square of a frequency in the sense of mechanics and the loss of stability occurs at $\lambda = 0$, i.e., at the transition from one equilibrium position into another nontrivial equilibrium position, *divergence* is said to occur. On the other hand, in Figure 8b, the two branches coalesce at $g_{1,\text{crit}}$, disappear, reappear at $g_{2,\text{crit}}$, and follow again separate traces from thereon. Hence, there exists a g interval between $g_{1\,\text{crit}}$ and $g_{2\,\text{crit}}$ for which λ is *complex*. Since for stability λ must be (positive) and real, as already mentioned, an instability domain is determined by $g_{1\,\text{crit}}$, $g_{2\,\text{crit}}$. However, this time a nonzero value of λ corresponds to the limiting values of g. For example, λ_{F1} in Figure 8b corresponds to $g_{1\,\text{crit}}$. Consequently, the mechanical system is not in an equilibrium position when instability occurs, contrary to the previous case, and the system becomes unstable by performing a so called *flutter* vibration with unboundedly increasing amplitudes.

The amplitudes λ can be evaluated approximately by means of the previously described method of Ritz. This method also yields relationship (2.21) and can be used to prove that for conservative systems always the case shown in Figure 8a is valid. Hence, *conservative systems are divergence type systems*. That can be seen in the following way: For the calculation of λ serves (2.20). Due to (2.18), λ is obtained from

$$\det(a_{ij} - \lambda\delta_{ij}) = 0. \tag{2.22}$$

If the matrix (a_{ij}) is symbolically denoted by

$$A = (a_{ij}), \tag{2.23}$$

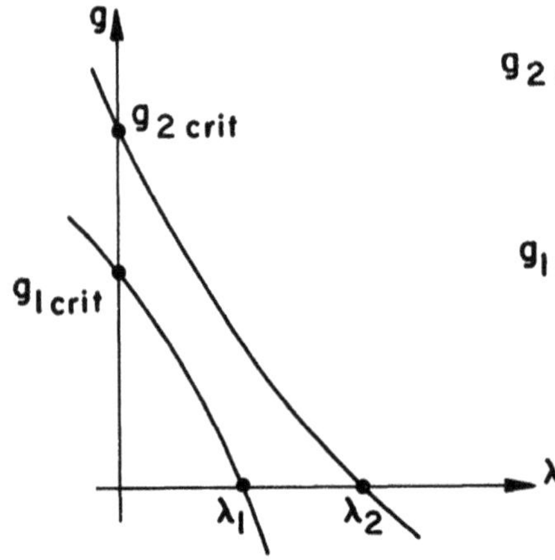

Figure 8a. Divergence type system

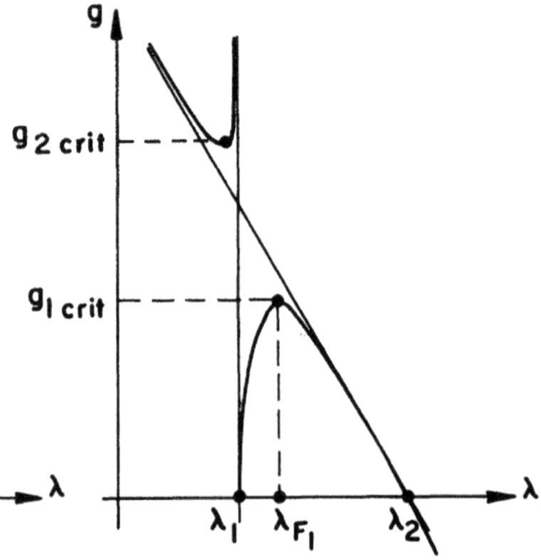

Figure 8b. Flutter type system

and if it is observed that (δ_{ij}) is the *unit matrix*

$$I = (\delta_{ij}), \tag{2.24}$$

then, it follows from (2.22) that λ is the *eigenvalue of the eigenvalue problem*

$$A - \lambda I = 0. \tag{2.25}$$

In (2.25), A and I are actually *infinite matrices*. For a numerical solution, the *method of reduction* must be applied to (2.25), i.e., one has to delete in A and I so many rows and columns that a quadratic $n \times n$ matrix remains. Then (2.25) changes into

$$A_n - \lambda_n I_n = 0, \tag{2.26}$$

which yields the approximation λ_n. Under the assumption that $\lambda_n \to \lambda$ for $n \to \infty$, conclusions may be drawn for λ from statements available for λ_n. The justification for such an approach will be given later when the convergence of λ_n towards λ will be proved.

The elements of matrix A_n are the expressions

$$a_{ij} = \int_B \sum_{k=0}^{2} h_k \, \phi_i^{(k)} \phi_j^{(k)} \, dB \tag{2.27}$$

as given by (2.16). It is obvious that $a_{ij} = a_{ji}$, i.e., that the matrix A_n is symmetric. But then, the eigenvalues λ_n of (2.26) are all real according to a well known theorem of matrix theory [7]. In the limit, λ is always real. Hence, instability can only occur through λ changing from positive to negative values. Therefore, only a behaviour of λ as shown in Figure 8a is possible, which means that the mechanical system is a divergence type system. Indeed, λ is never complex. However, the condition for flutter is that λ becomes complex. Thus flutter is ruled out.

Having determined λ, the evaluation of the solution w (eigenfunction corresponding to eigenvalue λ) can be tackled. For this purpose, λ is substituted in the algebraic system of equations (2.17) which then can be solved for the coefficients a_i. When these are known, they can be used together with λ to construct the solution w by means of (1.108), (1.109) and (1.110). Thereby one has reached the goal of the calculation.

Since the knowledge of λ is obviously the key to the mastership of the whole problem, more shall be said about the properties of λ and the convergence of the method of Ritz for the calculation of λ.

Assuming that all boundary conditions are satisfied, and taking into consideration that conditions $Q_t\,[w] = 0$, $Q_x\,[w] = 0$, $P\,[w] = 0$ are supposed to hold, the principle of Galerkin derived from (2.1) is obtained according to (1.47) in the form

$$\int_T \int_B \left\{ \frac{d}{dt}\left(\frac{\partial \mathscr{T}}{\partial w_t}\right) + \frac{\delta U}{\delta w} \right\} \delta w \, dB \, dt = 0. \tag{2.28}$$

Since δw is arbitrary, and because of the already mentioned boundary conditions, one is led corresponding to Section 1.3 and 1.4 to the boundary value problem

$$\frac{d}{dt}\left(\frac{\partial \mathscr{T}}{\partial w_t}\right) + \frac{\delta U}{\delta w} = 0, \qquad (Gw)_R = 0. \tag{2.29}$$

Using (1.42) in (2.4) yields

$$\frac{\delta U}{\delta w} = \sum_{k=0}^{2} (-1)^k \left[h_k(x)\, w^{(k)}(x,t)\right]^{(k)}. \tag{2.30}$$

By virtue of this relationship and by means of (2.2), equation (2.29) changes into

$$Tw + \sum_{k=0}^{2} (-1)^k \left[h_k\, w^{(k)}\right]^{(k)} = 0, \tag{2.31}$$

$$(Gw)_R = 0.$$

Using (1.100), the boundary value problem (2.31) can be rewritten to yield

$$(Tf)\, y + f \sum_{k=0}^{2} (-1)^k \, [h_k y^{(k)}]^{(k)} = 0,$$

$$f(Gy)_R = 0. \tag{2.32}$$

Dividing by f and using (1.101), problem (2.32) changes finally into

$$-\lambda y + My = 0, \; My = \sum_{k=0}^{2} (-1)^k \, [h_k y^{(k)}]^{(k)},$$

$$(Gy)_R = 0. \tag{2.33}$$

Operator M in (2.33) is formally selfadjoint as pointed out by E. Kamke in [8]. We shall require more of M, namely that it be *selfadjoint and positive definite even with respect to boundary conditions* $(Gy)_R = 0$. That is the case if

$$\int_B u\,(Mv)\,dB = \int_B v\,(Mu)\,dB, \tag{2.34}$$

$$\int_B u\,(Mu)\,dB > 0 \tag{2.35}$$

hold true for arbitrary *'comparison functions'* u, v. Condition (2.34) ensures selfadjointness, condition (2.35) ensures the positive definiteness of M. Comparison functions are functions which are often enough differentiable and which do satisfy *all* boundary conditions in (2.33).

Usually, My can be represented as

$$My = S_1\, y + g S_2 y, \tag{2.36}$$

where

$$S_1 y = h_0 y + [h_2 y'']'', \qquad S_2 y = -\,[\overset{*}{h}_1 y']', \tag{2.37}$$

$$\overset{*}{h}_1 = \frac{h_1}{g}\;.$$

Quantity g is the load parameter mentioned in connection with (2.15). Naturally, S_1 and S_2 are selfadjoint with respect to the boundary conditions since M is supposed to have this property.

From

$$-\lambda y + My = 0$$

follows

$$E = \int_B [-\lambda^2 + yS_1y + gyS_2y]\, dB = 0 \tag{2.38}$$

for the solution y, taking (2.36) into account. Functional E is an *energy expression* in the sense of mechanics. This functional yields for λ easily *Rayleigh's quotient.*

$$\lambda = \frac{\int_B [yS_1y + gyS_2y]dB}{\int_B y^2\, dB}\ . \tag{2.39}$$

By virtue of the selfadjointness of S_1 and S_2,

$$\int_B y\,(S_1\,\delta y)\, dB = \int_B \delta y\,(S_1 y)\, dB,$$

$$\int_B y\,(S_2\,\delta y)\, dB = \int_B \delta y\,(S_2 y)\, dB$$

according to (2.34). Hence, a variation of E yields according to (2.38)

$$-\delta\lambda \int_B y^2\, dB + \delta g \int_B yS_2y\, dB + 2\int_B [-\lambda y + S_1y + gS_2y]\delta y\, dB = 0. \tag{2.40}$$

Because of (2.36) and (2.33) the last term in (2.40) is equal to zero. Therefore, (2.40) can be transformed into

$$\frac{\delta g}{\delta\lambda} = \frac{\int_B y^2\, dB}{\int_B yS_2y\, dB}\ . \tag{2.41}$$

This quotient is obviously an expression of the slope of the eigenvalue curve (1.105) in a (g, λ) – axis system. As shown in Figure 8, flutter occurs when the (g, λ) – curve has a horizontal tangent. This condition reads analytically

$$\frac{\delta g}{\delta\lambda} = 0. \tag{2.42}$$

A glance at (2.41) shows that due to the nature of the numerator, condition (2.42) can never be satisfied. That is a second proof for the fact that conservative systems can never be flutter systems.

Reconsider (2.40) and assume that only λ and y are varied, but not g. Then, $\delta g \equiv 0$, and in (2.40) not only the last but also the last but one term are equal to zero. Hence,

$$\delta\lambda \int_B y^2 dB = 0 \tag{2.43}$$

remains of (2.40). Since the integral is never zero,

$$\delta\lambda = 0 \tag{2.44}$$

most hold true, which leads immediately to

$$\lambda = \text{stationary}.$$

Assuming in addition that the operator M is positive definite, so that (2.35) is valid, then $\lambda > 0$ because of (2.36), (2.39), and for the varied quantity $\bar{\lambda}$ the inequality

$$\bar{\lambda} = \frac{\int_B [uS_1 u + guS_2 u]dx}{\int_B u^2 dB} > 0 \tag{2.45}$$

holds true, likewise. But then the smallest, stationary value of the quantities $\bar{\lambda}$ must even be a minimum, i,e.,

$$\lambda_{min} = \min \bar{\lambda}. \tag{2.46}$$

This fact can be expressed in the following theorem: *The smallest eigenvalue λ of (2.33) is the minimum of the Rayleigh quotient calculated using arbitrary comparison functions u, if the operator M in (2.33) is selfadjoint and positive definite.*

As will be shown subsequently, it is sufficient to use '*admissible functions*' $\bar{\lambda}$ only instead of comparison functions. Admissible functions are functions which are sufficiently often continuously differentiable and which *satisfy the geometric boundary conditions only*. In order to make that apparent, the following proof of convergence will be given.

Take as an example the rod problem

$$-\lambda y + \alpha y^{IV} + [g(l - x)y']' = 0, \tag{2.47}$$

where α is the bending stiffness, g a uniformly distributed, conservative, compressive load, and l the length of the rod. In addition, the boundary conditions

$$y(0) = y''(0) = y(l) = y''(l) = 0 \tag{2.48}$$

for the simply supported rod,

$$y(0) = y'(0) = y(l) = y'(l) = 0 \tag{2.49}$$

for the clamped-clamped rod,

$$y(0) = y'(0) = y(l) = y''(l) = 0 \tag{2.50}$$

for the clamped-pinned rod,

$$y(0) = y'(0) = y''(l) = y'''(l) = 0 \tag{2.51}$$

for the clamped-free rod have to be considered.

For the *true solution* y of the problem, quotient (2.39) can be transformed by means of integration by parts into

$$\lambda = \frac{\int_0^l [\alpha y''^2 - g(l-x)y'^2]dx}{\int_0^l y^2 dx}, \tag{2.52}$$

since y satisfies all boundary conditions. The assertion is that (2.46) is correct also for

$$\bar{\lambda} = \frac{\int_0^l [\alpha \bar{y}''^2 - g(l-x)\bar{y}'^2]dx}{\int_0^l \bar{y}^2 dx}, \tag{2.53}$$

where $\bar{y}$ is only *admissible*. Then,

$$\bar{y} = y + \eta, \tag{2.54}$$

where η is an arbitrary, sufficiently often continuously differentiable function which satisfies the geometric boundary conditions only. Therefore, $\bar{y}$ satisfies likewise only the geometric but not all the boundary conditions.

Now, let the concept of the *norm* of a function be introduced. Let $h(x)$ be an L_2 – function, i.e., a function for which the integral

$$\int_0^l h^2 dx < \infty \tag{2.55}$$

exists and does have a finite value. Then, the quantity,

$$\| h(x) \| = \left[\int_0^l h^2 dx \right]^{1/2} \tag{2.56}$$

is said to be the norm of $h(x)$. Using the concept of the norm, the following *auxiliary theorem* can be formulated:

For arbitrary functions $\bar{y}$, which are sufficiently often differentiable, and which satisfy the geometric boundary conditions in (2.48) up to (2.51) (admissible functions), the inequalities

$$\| \bar{y} \|^2 \leq \frac{l^4}{4} \| \bar{y}'' \|^2, \quad \| \bar{y}' \|^2 \leq \frac{l^2}{2} \| \bar{y}'' \|^2 \tag{2.57}$$

hold true.

Proof: Obviously,

$$\bar{y}'(x) - y'(0) = \int_0^x \bar{y}''(x)\, dx. \tag{2.58}$$

Hence,

$$[\bar{y}'(x) - \bar{y}'(0)]^2 = \left[\int_0^x \bar{y}'' dx \right]^2 \leq x \int_0^x \bar{y}''^2 dx \leq x \int_0^l \bar{y}''^2 dx \tag{2.59}$$

$$= x \| \bar{y}'' \|^2$$

according to the *theorem of Schwarz* [9]. Moreover,

$$\int_0^l [\bar{y}'(x) - \bar{y}'(0)]^2 dx = \int_0^l \bar{y}'^2 dx - 2\bar{y}'(0) \int_0^l \bar{y}'(x) dx + [\bar{y}'(0)]^2 l,$$

$$\int_0^l [\bar{y}'(x) - \bar{y}'(0)]^2 dx = \| \bar{y}' \|^2 - 2\bar{y}'(0)\,[\bar{y}(l) - \bar{y}(0)] + [\bar{y}'(0)]^2 l, \tag{2.60}$$

using (2.56). Due to boundary conditions (2.48) to (2.51), the second term on the right hand side of (2.60) vanishes in any case. Therefore,

$$\int_0^l [\bar{y}'(x) - \bar{y}'(0)]^2 dx = \| \bar{y}' \|^2 + [\bar{y}'(0)]^2 l. \tag{2.61}$$

Using (2.59) in (2.61) yields

$$\|\bar{y}'\|^2 \leqq \frac{l^2}{2} \|\bar{y}''\|^2, \qquad (2.62) \text{ or } (2.57_2)$$

which is the second statement in the auxiliary theorem. Since always $\bar{y}(0) = 0$ due to the boundary conditions, also

$$\bar{y}(x) = \int_0^x \bar{y}'(x)\,dx$$

holds true. Hence, using the inequality of Schwarz,

$$\bar{y}^2(x) = \left[\int_0^x \bar{y}'(x)dx\right]^2 \leqq x\int_0^x \bar{y}'^2 dx \leqq x\int_0^l \bar{y}'^2 dx = x\|y'\|^2. \qquad (2.63)$$

But then

$$\int_0^l \bar{y}^2 dx \leqq \frac{l^2}{2} \|\bar{y}'\|^2, \quad \|\bar{y}\|^2 \leqq \frac{l^2}{2} \|\bar{y}'\|^2. \qquad (2.64)$$

Combining (2.64) with (2.57_2) yields finally

$$\|\bar{y}\|^2 \leqq \frac{l^4}{4} \|\bar{y}''\|^2. \qquad (2.57_1)$$

That is the first statement in the auxiliary theorem.

The auxiliary theorem applies to admissible functions $\bar{y}$. But then it applies even more to comparison functions u and to the solution y. Using it for an estimate of the functional

$$F = \int_0^l [\alpha\bar{y}''^2 - g(l-x)\bar{y}'^2]dx,$$

the result is

$$F \geqq \left(\alpha - g\frac{l^3}{2}\right) \|\bar{y}''\|^2. \qquad (2.65)$$

This inequality holds also true, if F is written down not for an admissible function $\bar{y}$ but for a comparison function u for the solution y.

Due to (2.65) and (2.57_1),

$$\lambda, \bar{\lambda} > \left(\alpha - g\,\frac{l^3}{2}\right)\frac{4}{l^4} \tag{2.66}$$

follows from (2.52) and (2.53).

For comparison functions u,

$$\int_0^l u(Mu)dx = \int_0^l \{\alpha u^{\mathrm{IV}} u + g[(l-x)u']'u\}dx = \int_0^l [\alpha u''^2 - g(l-x)u'^2]dx,$$

which is obtained by integration by parts observing the boundary conditions. Because of this relationship and because of (2.65), the inequality

$$\int_0^l u(Mu)\,dx \geqq \left(\alpha - g\,\frac{l^3}{2}\right) \| u'' \|^2 \tag{2.67}$$

is obtained.

After these preparations, the very proof of convergence of the method of Ritz can be tackled, referring to the adopted example. The proof will be carried out under the following *assumptions:*

A: $\alpha - g\,\dfrac{l^3}{2} > 0.$ (2.68)

B: The solution y of the problem be continuous and sufficiently often continuously differentiable so that

$$|\,y^{(k)} - \sum_{i=1}^{m} a_i \phi_i^{(k)}| \leqq \epsilon, \quad k = 0,1,2, \tag{2.69}$$

for arbitrary small, positive ϵ and for properly chosen m. Here, the a_i are generalized Fourier coefficients and the ϕ_i are elements of a *complete* system of functions [10].

With respect to the assumptions, the following remarks may be made:

A: a) Because of (2.66), condition (2.68) yields

$$\lambda > 0, \bar{\lambda} > 0; \tag{2.70}$$

b) because of (2.65) and (2.67), the functional F and the operator M are positive definite. This is true, since for $\bar{y}'' \neq 0, u'' \neq 0$, also $F > 0$, $\int_0^l u\,(Mu)\,dx > 0$, and for $\bar{y}'' \equiv 0, u'' \equiv 0$ follows $y \equiv 0, u \equiv 0$ due to the geometric boundary conditions so that indeed only then $F = 0$, $\int_0^l u\,(Mu)\,dx = 0$, as it is supposed to be in the case of positive definiteness.

In Figure 9, condition (2.68) is graphically portrayed. It is obvious that at least a part of the eigenvalue curve of the mechanical system serving as an example is covered by the proof of convergence.

B: Condition (2.69) can be satisfied for example if the ϕ_i are chosen as appropriate polynomials in x. This follows from the *approximation theorem of Weierstrass* and from additional statements by Courant-Hilbert [11].

For the proof of convergence, consider first the variation of (2.53), which reads

$$\delta\bar{\lambda} = 2\,\frac{\int_0^l [\alpha\bar{y}''\delta\bar{y}'' - g(l-x)\bar{y}'\delta\bar{y}' - \lambda\bar{y}\delta\bar{y}]\,dx}{\int_0^l \bar{y}^2 dx}\,. \qquad (2.71)$$

Chosing in (2.71) especially

$$\bar{y}^* = y, \bar{\lambda}^* = \lambda,$$

i.e., considering the variation (2.71) at the point in the function space specified by the solution y and the eigenvalue λ of (2.47), yields

$$\delta\lambda = 2\,\frac{\int_0^l [\alpha y''\eta'' - g(l-x)y'\eta' - \lambda y\eta]\,dx}{\int_0^l y^2 dx}\,, \qquad (2.72)$$

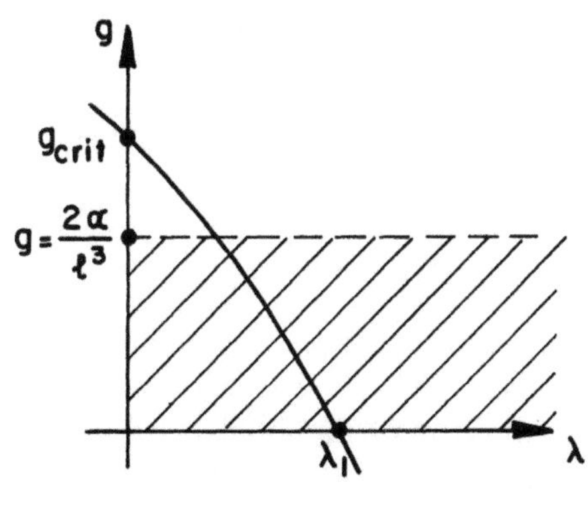

Figure 9

since $\bar{y} = y + \eta = y + \delta y$, and $\delta y = \eta$, where η is an admissible function only.

Integrating the numerator of (2.72) by parts yields

$$\int_0^l [\alpha y'' \eta'' - g(l-x) y' \eta' - \lambda y \eta] dx = \int_0^l \{\alpha y^{IV} + g[(l-x)y']'$$

$$-\lambda y\} \eta \, dx, \tag{2.73}$$

although η is an admissible function satisfying the geometric boundary conditions only. This result justifies the use of only admissible functions in the method of Ritz.

Using (2.73) in (2.72) yields

$$\delta\lambda = 2 \frac{\int_0^l \{\alpha y^{IV} + g[(l-x)y']' - \lambda y\} \eta \, dx}{\int_0^l y^2 dx} \tag{2.74}$$

Because of (2.47), again,

$$\delta\lambda = 0 \tag{2.44}$$

follows from (2.74). The quantity $\bar{\lambda}$ has therefore at least a stationary value at the point λ. Due to (2.68), the *smallest* eigenvalue λ_{min} of (2.47) must even be a minimum. Hence,

$$\lambda_{min} = \min \bar{\lambda} \tag{2.75}$$

in the *space of functions which are at least admissible*. But then,

$$\bar{\lambda} \geqq \lambda_{min}. \tag{2.76}$$

This statement completes the first part of the proof.

Function

$$y_n = \sum_{k=1}^{n} a_k^{(n)} \phi_k \tag{2.77}$$

belongs to the class of admissible functions if the ϕ_k are elements of a complete system of functions which satisfy the geometric boundary conditions. In the second part of the proof it will be shown that the y_n can be determined and therefore do exist. For this purpose, the expression

$$\bar{\lambda}(y_n) = \frac{\int_0^l \left\{ \alpha \left[\sum_{k=1}^{n} a_k^{(n)} \phi_k'' \right]^2 - g(l-x) \left[\sum_{k=1}^{n} a_k^{(n)} \phi_k' \right]^2 \right\} dx}{\int_0^l \left[\sum_{k=1}^{n} a_k^{(n)} \phi_k \right]^2 dx} \tag{2.78}$$

is calculated. If y_n exists, condition (2.76) applies to (2.78), and one is led as near as possible to λ, if one selects the coefficients $a_k^{(n)}$ in (2.78) such that $\bar{\lambda}\,(y_n)$ assumes its minimum. This is the actual philosophy behind the method of Ritz. It leads to the condition

$$\frac{\partial \bar{\lambda}}{\partial a_i} = 0. \tag{2.79}$$

Applying (2.79) to (2.78) yields after some elementary calculations

$$\int_0^l \left\{ \alpha \left[\sum_{k=1}^{n} a_k^{(n)} \phi_k'' \right] \phi_i'' - g(l-x) \left[\sum_{k=1}^{n} a_k^{(n)} \phi_k' \right] \phi_i' \right\} dx - \bar{\lambda} \int_0^l \left[\sum_{k=1}^{n} a_k^{(n)} \phi_k \right] \phi_i \, dx = 0.$$

Assuming that the system of functions ϕ_i is orthonormal, which does not restrict generality, one can transform further to yield

$$\sum_{k=1}^{n} a_k^{(n)} \left\{ -\bar{\lambda}\delta_{ik} + \int_0^l [\alpha \phi_k'' \phi_i'' - g(l-x)\phi_k' \phi_i']\,dx \right\} = 0. \tag{2.80}$$

This equation corresponds to (2.14). That equation had been obtained by applying (1.111) to (2.1), (2.8), respectively. The conclusion is that the proof of convergence given here does apply to the method of Ritz as outlined on pages 30 to 36.

Introducing

$$\rho_{ik} = -\bar{\lambda}\delta_{ik} + \int_0^l [\alpha \phi_k'' \phi_i'' - g(l-x)\phi_k' \phi_i']\,dx, \tag{2.81}$$

and using it in (2.80), yields

$$\sum_{k=1}^{n} a_k^{(n)} \rho_{ik} = 0, \tag{2.82}$$

which can be used to determine the coefficients $a_k^{(n)}$. These coefficients exist, and hence functions y_n exist according to (2.77) if $\bar{\lambda}$ is chosen as the eigenvalue of

$$\det \rho_{ik} = 0. \tag{2.83}$$

But (2.83) yields certainly an eigenvalue $\bar{\lambda}(y_n)$. Hence, functions y_n do exist.

In the third and last part of the proof it will be shown that indeed

$$\bar{\lambda}(y_n) \to \lambda \text{ for } n \to \infty. \tag{2.84}$$

Since $\bar{\lambda}(y_n)$ is calculated by means of (2.83), and since (2.83) is identical to (2.22), it follows again, as already mentioned, that the proof of convergence covers that variety of the method of Ritz described at the beginning of this section.

Let z be a function in the class defined by (2.77). The existence of this class has just been proved. Now, it is claimed that

$$|\bar{\lambda}(z) - \lambda| \leqq \delta. \tag{2.85}$$

In order to prove this, the expression

$$\bar{\lambda}(z) - \lambda = \frac{\int\limits_0^l [\alpha z''^2 - g(l-x)z'^2]dx}{\int\limits_0^l z^2 dx} - \frac{\int\limits_0^l [\alpha y''^2 - g(l-x)y'^2]dx}{\int\limits_0^l y^2 dx}$$

has to be estimated. Also,

$$\bar{\lambda}(z) - \lambda = \frac{\left[\int\limits_0^l (y^2 - z^2)dx\right]\left\{\int\limits_0^l [\alpha z''^2 - g(l-x)z'^2]dx\right\}}{\left[\int\limits_0^l z^2 dx\right]\left[\int\limits_0^l y^2 dx\right]}$$

$$+ \frac{\left[\int\limits_0^l z^2 dx\right]\left\{\int\limits_0^l [\alpha(z''^2 - y''^2) - g(l-x)(z'^2 - y'^2)]dx\right\}}{\left[\int\limits_0^l z^2 dx\right]\left[\int\limits_0^l y^2 dx\right]}.$$

Hence,

$$|\bar{\lambda}(z)-\lambda| \leqq \tag{2.86}$$

$$\leqq \frac{1}{\left[\int_0^l z^2 dx\right]\left[\int_0^l y^2 dx\right]} \left\{\int_0^l [|y-z|\cdot|y+z|]dx\right\} \cdot \left|\left\{\int_0^l [\alpha z''^2\right.\right.$$

$$\left.\left.-g(l-x)z'^2]dx\right\}\right| + \left[\int_0^l z^2 dx\right]\left\{\int_0^l [|\alpha|\cdot|z''-y''|\cdot|z''+y''|\right.$$

$$\left.+g(l-x)|z'-y'|\cdot|z'+y'|]dx\right\}.$$

Due to the nature of z, condition (2.69) can be applied to the terms $|y^{(k)} -z^{(k)}|$, $k = 0,1,2$, in (2.86). Since all other quantities in (2.86) are bounded, one concludes immediately that

$$|\bar{\lambda}(z)-\lambda| \leqq \epsilon \cdot K, \quad K < \infty. \tag{2.87}$$

Using the notation $\delta = \epsilon \cdot K$, (2.87) becomes identical to (2.85). Hence, that inequality is in fact valid.

Let y_n and y_m be two functions in the class defined by (2.77) to which also z belongs. When $n \geqq m$, certainly

$$\bar{\lambda}(y_n) < \bar{\lambda}(y_m), \tag{2.88}$$

since y_n involves a larger number of parameters for minimizing $\bar{\lambda}(y_n)$ than y_m possesses for minimizing $\bar{\lambda}(y_m)$.

But y_m can be used to replace z in (2.85), since both functions are in the same class. Therefore also

$$|\bar{\lambda}(y_m)-\lambda| \leqq \delta. \tag{2.89}$$

Due to (2.88) and (2.70) one must conclude that

$$|\bar{\lambda}(y_n)-\lambda| < |\bar{\lambda}(y_m)-\lambda| \leqq \delta,$$

so that

$$|\bar{\lambda}(y_n)-\lambda| < \delta. \tag{2.90}$$

A glance on (2.69) shows that ϵ, and thus δ, tends to zero for $n \to \infty$. Therefore, the estimate (2.90) yields

$$\bar{\lambda}(y_n) \to \lambda \text{ for } n \to \infty, \tag{2.84}$$

which proves the convergence of the method of Ritz.

It may be mentioned in addition that the transition from (2.90) to (2.84) is possible because $\bar{\lambda}(y_n)$ and λ are positive by assumption, and because (2.76) holds true.

Up to now it has been especially emphasized that the coordinate functions ϕ_i of the method of Ritz used in the series (1.111) are allowed to be 'admissible functions' which satisfy the geometric boundary conditions only and not all of the boudary conditions. The proof of convergence has confirmed the correctness of this assertion. The possibility of working with expansions in terms of admissible functions is certainly of great practical value. One has only to realize that dynamic boundary conditions, which for problems of order $2m$ involve derivatives of order m and of an order higher than m [13], are complicated and not easily satisfied. If they can be avoided, the choice of coordinate functions is facilitated, thus perhaps enabling one at all to carry out the method of Ritz numerically.

On the other hand, the rate of convergence and thus the accuracy of the results is essentially improved if 'comparison functions' are used in the series (1.111) which satisfy *all* boundary conditions. If such comparison functions are available, they should have preference over any other kind of admissible functions. In Section 1.5 it has already been reported how a system of coordinate functions may be generated, and how these functions may be used for series expansions, for example according to a theorem of L. Collatz [6]. In the following, coordinate functions are therefore assumed to satisfy all boundary conditions. This assumption allows for a new proof of convergence. This proof is in the first place of interest because it can be generalized to apply in the nonclassical case which will be dealt with in the following sections.

The fundamental variational principle (2.1), which is the very foundation of the classical method of Ritz, is equivalent to the derived principle (2.28), if all boudary conditions are being satisfied. Hence, principle (2.28) may be the starting point for the subsequent investigations. By means of (2.2), (2.30), (1.100), (1.101), and the operator M defined in (2.33), principle (2.28) can be changed into

$$\int_0^l [My - \lambda y]dx = 0, \tag{2.91}$$

if the domain of integration B is properly defined, and if the integration with respect to t is suppressed.

If in the expansion (1.111) the coordinate functions ϕ_i do satisfy all boundary conditions, (1.111) can be used in (2.91). That leads to

$$\int_0^l [M(\sum_i c_i\phi_i)-\lambda(\sum_i c_i\phi_i)]\,\phi_j dx = 0 \quad j = 1,2,3, \ldots . \tag{2.92}$$

Assuming again that the system ϕ_i is orthonormal, and using the abbreviation

$$a_{ij} = \int_0^l (M\phi_i)\phi_j dx, \tag{2.93}$$

relationship (2.92) becomes the infinite system of equations

$$\sum_i c_i (a_{ij} - \lambda\delta_{ij}) = 0. \tag{2.94}$$

The system of Ritz-Galerkin equations is obtained from (2.94) by reducing that system into the finite system

$$\sum_{i=1}^{n} c_i(a_{ij} - \lambda_n \delta_{ij}) = 0, \quad j = 1,2,3, \ldots, n. \tag{2.95}$$

It remains to be proved that for the λ_n calculated by means of (2.95) the condition

$$\lambda_n \to \lambda \text{ for } n \to \infty$$

holds true.

For this proof assume that the functions ϕ_i are specifically the eigenfunctions of operator M with respect to the prescribed boundary conditions. Then,

$$\int_0^l (M\phi_i)\phi_j dx = \lambda_i \phi_{ij}, \tag{2.96}$$

where λ_i is the eigenvalue corresponding to eigenfunction ϕ_i. By means of (2.96), equation (2.92) changes into

$$c_j\lambda_j = \lambda \int_0^l (\sum_i c_i\phi_i)\,\phi_j dx. \tag{2.97}$$

In the sense of the method of Ritz, the sum on the right hand side of (2.97) is a finite one. Therefore,

$$\sum_{i=1}^{n} c_i \phi_i = y_n \tag{2.98}$$

may be used in the following. The result of this, and a simple transformation yields

$$c_j = \lambda \int_0^l y_n(x) \frac{\phi_j(x)}{\lambda_j} dx. \tag{2.99}$$

Multiplying both sides in (2.99) by $\phi_j(\xi)$ and adding with respect to j yields

$$\sum_{j=1}^{n} c_j \phi_j(\xi) = \lambda \int_0^l y_n(x) \sum_{j=1}^{n} \frac{\phi_j(x)\phi_j(\xi)}{\lambda_j} dx. \tag{2.100}$$

But for the selfadjoint operator M and the prescribed boundary conditions there exists a Green function $G(x, \xi)$ which admits of a bilinear expansion [14], which converges at least in the mean. Therefore the approximation

$$G_n(x, \xi) = \sum_{j=1}^{n} \frac{\phi_j(x)\phi_j(\xi)}{\lambda_j} \tag{2.101}$$

can be assumed to hold for $G(x, \xi)$. Using (2.101) and (2.98), equation (2.100) can be rewritten to yield

$$y_n(\xi) = \lambda \int_0^l G_n(x, \xi) y_n(x) dx. \tag{2.102}$$

Hence, it has been shown that the method of Ritz consists in replacing integral equation

$$y(\xi) = \lambda \int_0^l G(x, \xi) y(x) dx, \tag{2.103}$$

which is equivalent to the boundary value problem (2.33), by the approximation (2.102). In this approximation, not only the unknown function y, but also Green's function has been substituted by a series, as can be seen by comparing (2.100) with (2.98) and (2.101). This procedure corresponds completely to the *methods of moments* [15], as will be shown in Section 4.

The special system of eigenfunctions of M has been used in (2.92) only for the sake of being able to prove the close relationship between (2.100) and (2.103). In practical cases, these eigenfunctions will not be available. Because if they were, it would mean that the actual problem (2.33) were already solved. However, it is easy to change over to an arbitrary, complete, orthonormal system of comparison functions $\psi_i(x)$ in (2.100). This corresponds to a 'coordinate transformation in the function space' [16] which is defined by

$$\phi_k = \sum_s b_{ks} \psi_s. \tag{2.104}$$

Hence,

$$y_n(x) = \sum_{k=1}^{n} c_k \phi_k(x) = \sum_{k,s=1}^{n} c_k b_{ks} \psi_s(x) = \sum_{s=1}^{n} \alpha_s \psi_s(x), \tag{2.105}$$

where

$$\alpha_s = \sum_{k=1}^{n} c_k b_{ks} = \int_0^l y(x) \psi_s(x) dx < \infty. \tag{2.106}$$

Correspondingly,

$$\sum_r \frac{\phi_r(x)\phi_r(\xi)}{\lambda_r} = \sum_r \frac{\sum_s b_{rs} \psi_s(x) \sum_t b_{rt} \psi_t(\xi)}{\lambda_r} = \sum_{s,t} d_{st} \psi_s(x) \psi_t(\xi), \tag{2.107}$$

where

$$d_{st} = \sum_r \frac{b_{rs} b_{rt}}{\lambda_r} . \tag{2.108}$$

Using (2.105) and (2.107), equation (2.100) can be changed into

$$\sum_{s=1}^{n} \alpha_s \psi_s(\xi) = \lambda \int_0^l \left\{ \left[\sum_{m=1}^{n} \alpha_m \psi_m(x) \right] \left[\sum_{p,t=1}^{n} d_{pt} \psi_p(x) \psi_t(\xi) \right] \right\} dx. \tag{2.109}$$

Executing the integration on the right hand side of (2.109) and observing the orthonormality of the ψ_i yields

$$\sum_{s=1}^{n} \alpha_s \psi_s(\xi) = \lambda \sum_{m,t=1}^{n} \alpha_m d_{mt} \psi_t(\xi),$$

from which follows

$$\alpha_s = \lambda \sum_{m=1}^{n} \alpha_m d_{ms}, \qquad s = 1,2, \ldots, n, \tag{2.110}$$

In (2.111), λ has been replaced by λ_n because (2.111) yields certainly only an approximation of the exact eigenvalue λ.

The systems of equations (2.95) and (2.111) are equivalent. Obviously, (2.111) follows from (2.95) when writing $i = m$, $j = s$, $c = \alpha$ in the latter system and multiplying it by the inverse of the matrix (a_{ij}). The inverse $(a_{ij})^{-1}$ exists indeed, since (a_{ij}) is a *finite* matrix.

Changing from (2.95) to (2.111) has the advantage that for (2.111) a proof of convergence is easy to find: For evaluating the exact eigenvalue λ, equation (2.103) may be used. The solution y can be expanded into the absolutely and uniformly convergent series [6]

$$y = \sum_{k=1}^{\infty} \alpha_k \psi_k. \tag{2.112}$$

Also,

$$G(x, \xi) = \sum_{s,t=1}^{\infty} d_{st} \psi_s(x) \psi_t(\xi) \tag{2.113}$$

holds true if $G(x, \xi)$ is symmetric, continuous, and if the eigenvalues of the problem are positive with the exception of a finite number of them [17]. By means of (2.112), (2.113), the integral equation (2.103) changes after some calculation into

$$\sum_{m=1}^{\infty} \alpha_m (\delta_{ms} - \lambda d_{ms}) = 0, \qquad s = 1,2, \ldots, \infty. \tag{2.114}$$

Since this system of equations is a homogeneous one, the exact eigenvalue λ is obtained from

$$\det(\delta_{ms} - \lambda d_{ms}) = 0, \qquad m, s = 1,2, \ldots, \infty, \tag{2.115}$$

where the determinant is an infinite one. The approximation following from the method of Ritz was (2.111), yielding for the approximation λ_n of the eigenvalue the condition

$$\det(\delta_{ms} - \lambda_n d_{ms}^{(n)}) = 0, \qquad m, s = 1,2, \ldots, n. \tag{2.116}$$

The determinant in (2.116) is a *finite* $n \times n$ determinant. Obviously, the finite system of equations (2.111) is obtained from the infinite system of equations (2.114) by applying the *method of reduction*. The finite system is an arbitrarily close approximation to the infinite system [18] if λ_n is finite, and if

$$\sum_{m,s=1}^{\infty} | d_{ms} |^2 < \infty \tag{2.117}$$

holds true. However, because of (2.113),

$$\sum_{m,s=1}^{\infty} | d_{ms} |^2 = \int_0^l \int_0^l G^2(x, \xi) dx\, d\xi < \infty, \tag{2.118}$$

hence, (2.117) is satisfied if $G(x, \xi)$ is square integrable.

Moreover,

$\lambda_n \to \lambda$ for $n \to \infty$

if $\det(\delta_{ms} - \lambda d_{ms}^{(n)})$ tends toward $\det(\delta_{ms} - \lambda d_{ms})$ for increasing n. That is according to H. von Koch [19] under the assumption of λ, respectively λ_n, being finite the case, if in addition to (2.117)

$$\sum_{i=1}^{\infty} | d_{ii} | < \infty \tag{2.119}$$

holds true. Under the assumption that $G(x, \xi)$ is positive definite [20],

$$\int_0^l \int_0^l G(x, \xi) \psi_i(x) \psi_i(\xi) dx\, d\xi = d_{ii} > 0, \tag{2.120}$$

making use of (2.113). Also, (2.113) leads after a simple calculation to

$$\int_0^l G(x, x) dx = \sum_{i=1}^{\infty} d_{ii}. \tag{2.121}$$

Relationships (2.121) and (2.120) yield immediately (2.119) if $G(x, x)$ is integrable. Consequently, the following theorem results:

If Green's function of the posed problem is positive definite, integrable, square integrable and expandable in the form (2.113), than the value λ_n furnished by the method of Ritz converges toward the exact eigenvalue λ if λ is finite.

In the majority of the cases, Green's function of the problem is not known explicitely. Because if it were, then more powerful methods than the method of Ritz could be used, for example, the method of Grammel (see Section 4.3). But if the operator M in (2.33) is selfadjoint and positive definite, then there exists a symmetric, positive definite, continuous and thus integrable and square integrable Green function, which in addition is bilinearly expandable [21]. For problems of this kind the conditions of the above mentioned theorem are therefore satisfied in advance. Hence, for these problems, the method of Ritz does converge.

2.2 The general, nonconservative case

This time assume that the nonconservative forces $Q_x[w]$ and $P[w]$ are present. Then, the fundamental variational principle (1.19) reads

$$\int_T \left[\int_B \left\{ \frac{d}{dt}\left(\frac{\partial \mathscr{T}}{\partial w_t}\right)\delta w + \delta \mathscr{U} - Q_x[w]\delta w \right\} dB - \int_{R_F} P[w]\delta w \, dR \right] dt = 0. \tag{2.122}$$

Contrary to (1.19), it has already been taken into account in (2.122) that $\delta w \neq 0$ holds true only along the free edge R_F so that $P[w]\,\delta w$ is different from zero only there, and not along the whole edge R.

Using in (2.122)

$$\frac{d}{dt}\left(\frac{\partial \mathscr{T}}{\partial w_t}\right) = Tw, \tag{2.2}$$

$$\mathscr{U} = \frac{1}{2}\sum_{k=0}^{2} h_k(x)[w^{(k)}(x,t)]^2, \tag{2.4}$$

$$w(x,t) = f(t)y(x), \tag{1.100}$$

and according to (1.56)

$$Q_x[w] = -gw_x, \tag{2.123}$$

$$P[w] = -Pw_x, \tag{2.124}$$

where g and P are load parameters, yields

$$\int_T f^2\, dt \left[\int_B \left\{ \frac{Tf}{f} y\delta y + \delta\left(\frac{1}{2}\sum_{k=0}^{2} h_k(x)[y^{(k)}(x)]^2\right) + gy'\delta y \right\} dB + \int_{R_F} Py'\delta y \, dR \right] = 0. \tag{2.125}$$

By means of (1.101), equation (2.125) can be changed into

$$\int_T f^2\, dt \left[\int_B \left\{ -\lambda y\delta y + \delta\left(\frac{1}{2}\sum_{k=0}^{2} h_k(x)[y^{(k)}(x)]^2\right) + gy'\delta y \right\} dB + \int_{R_F} Py'\delta y \, dR \right] = 0. \tag{2.126}$$

This condition can only be satisfied if

$$\int_B \left\{ -\lambda y\delta y + \delta\left(\frac{1}{2}\sum_{k=0}^{2} h_k(x)[y^{(k)}(x)]^2\right) + gy'\delta y \right\} dB + \int_{R_F} Py'\delta y \, dR = 0. \tag{2.127}$$

Consequently,

$$\delta\left[\int_B \frac{1}{2}\left\{-\lambda y^2 + \sum_{k=0}^{2} h_k(x)[y^{(k)}(x)]^2\right\} dB\right]$$

$$= -\left[\int_B gy'\delta y\, dB + \int_{R_F} Py'\delta y\, dR\right], \tag{2.128}$$

which shows that in the presence of nonconservative forces the fundamental variational principle is no longer an extremum principle.

Now, the method of Ritz will be applied to

$$\int_B \left\{-\lambda y\delta y + \sum_{k=0}^{2} h_k y^{(k)}\delta y^{(k)} + gy'\delta y\right\} dB + \int_{R_F} Py'\delta y\, dR = 0, \tag{2.129}$$

a relationship which follows from (2.127), (2.128), respectively. For this purpose the series

$$y(x) = \sum_i a_i\phi_i(x) \tag{1.111}$$

will be used in which the coordinate functions $\phi_i(x)$ satisfy all the boundary conditions or at least the geometric ones. Once again

$$y^{(k)} = \sum_i a_i\phi_i^{(k)},\ \delta y^{(k)} = \frac{\partial y^{(k)}}{\partial a_i}\,\delta a_i = \phi_i^{(k)}\delta a_i, \tag{2.11}$$

which is to be used in (2.129) together with (1.111). The result is

$$\int_B \Big\{-\lambda(\sum_i a_i\phi_i)(\sum_j \phi_j\delta a_j) + \sum_{k=0}^{2} h_k(\sum_i a_i\phi_i^{(k)})(\sum_j \phi_j^{(k)}\delta a_j)$$

$$+ g(\sum_i a_i\phi_i')(\sum_j \phi_j\delta a_j)\Big\} dB + \int_{R_F} P(\sum_i a_i\phi_i')(\sum_j \phi_j\delta a_j)\, dR = 0, \tag{2.130}$$

which yields

$$\int_B \left\{-\lambda(\sum_i a_i\phi_i)\phi_j + \sum_{k=0}^{2} h_k(\sum_i a_i\phi_i^{(k)})\phi_j^{(k)} + g(\sum_i a_i\phi_i')\phi_j\right\} dB$$

$$+ \int_{R_F} P(\sum_i a_i\phi_i')\phi_j\, dR = 0, \qquad j = 1,2,3,\ldots, \tag{2.131}$$

due to the arbitrariness of the variations δa_j. Because of (1.112), i.e., because of the orthonormality of the coordinate functions, and because of the additivity of the integration, (2.131) changes into

$$\sum_i a_i \left\{ -\lambda\delta_{ij} + \int_B \left[\sum_{k=0}^{2} h_k \phi_i^{(k)} \phi_j^{(k)} + g\phi_i' \phi_j \right] dB + \int_{R_F} P\phi_i' \phi_j \, dR \right\} = 0. \quad (2.132)$$

Like in Section 2.1 assume

$$h_1(x) = g h_1^*(x). \quad (2.15)$$

Then,

$$\int_B \left[\sum_{k=0}^{2} h_k \phi_i^{(k)} \phi_j^{(k)} + g\phi_i' \phi_j \right] dB + \int_{R_F} P\phi_i' \phi_j \, dR = \tilde{a}_{ij}(g, P), \quad (2.133)$$

and (2.132) can be changed into

$$\sum_i a_i \left\{ -\lambda\delta_{ij} + \tilde{a}_{ij}(g, P) \right\} = 0, \quad j = 1,2,3, \ldots . \quad (2.134)$$

Using

$$\tilde{\rho}_{ij} = \tilde{a}_{ij}(g, P) - \lambda\delta_{ij} \quad (2.135)$$

as an abbreviation, it becomes apparent that (2.134) is the linear, algebraic system

$$\sum_i a_i \tilde{\rho}_{ij} = 0. \quad (2.136)$$

This system yields nontrivial solutions for the coefficients a_i in (1.111) only if

$$\det \tilde{\rho}_{ij} = 0. \quad (2.137)$$

Since the $\tilde{\rho}_{ij}$ are functions of g, P, and λ according to (2.135), the method of Ritz leads by means of (2.137) finally to the eigenvalue equation

$$\tilde{F}(g, P, \lambda) = 0. \quad (2.138)$$

This equation yields

$$\lambda = \lambda(g, P). \quad (2.139)$$

Hence, the eigenvalue λ is a function of the load parameters g and P, which therefore through their influence on λ exert an influence on the stability behaviour of the system. For the sake of simplicity it shall be assumed in the following that parameter P is absent in the problems to be considered subsequently. Then, λ depends on g, and eigenvalue curves are obtained which have been already discussed in Section 2.1 and shown in Figure 8.

Writing down equation (2.137) in the form

$$\det (\tilde{a}_{ij} - \lambda \delta_{ij}) = 0 \tag{2.140}$$

proves that one has to deal essentially with the eigenvalue problem of the "structural matrix"

$$\tilde{A} = (\tilde{a}_{ij}). \tag{2.141}$$

Since according to (2.133)

$$\tilde{a}_{ij} \neq \tilde{a}_{ji}, \tag{2.142}$$

matrix $\tilde{A}$ is nonsymmetric, and its eigenvalues may be real as well as complex. Therefore, a trace of the eigenvalue curve as in Figure 8b is now possible, i.e., *systems with a nonconservative loading may possibly be flutter systems.*

Having calculated λ, the solution w, which is the eigenfunction corresponding to the eigenvalue λ, can be determined as indicated in Section 2.1. In order to proceed toward this goal, quantity λ, now assumed to be known, must be used in (2.134), (2.132), respectively, and then this homogeneous, algebraic system of equations be solved for the coefficients a_i. One of the coefficients remains however arbitrary due to the system being homogeneous. Once the a_i being known, relationship (1.111) yields $y(x)$. Function $f(t)$ follows from (1.108), (1.109) by means of the known λ. Finally, $w(x, t)$ follows from (1.100).

As already mentioned in the previous section, the whole process of calculation hinges essentially around λ. Also, the behaviour of λ will determine the nature of the solution and thus stability and instability of the system. For that reason, a further discussion of λ shall be undertaken in the following.

Assuming that the variations in (2.122) do *satisfy all boundary conditions,* one may work with Galerkin's principle (1.47) which was derived from (2.122). For $Q_t [w] \equiv 0$, which is assumed to hold true, it changes specifically into

$$\int_T \int_B \left\{ \frac{d}{dt} \left(\frac{\partial \mathcal{T}}{\partial w_t} \right) + \frac{\delta U}{\delta w} - Q_x [w] \right\} \delta w \, dB \, dt = 0. \tag{2.143}$$

Due to the arbitrariness of the δw, and using the boundary conditions, one is led to the boundary value problem

$$\frac{d}{dt}\left(\frac{\partial \mathscr{T}}{\partial w_t}\right) + \frac{\delta U}{\delta w} - Q_x[w] = 0, \tag{2.144}$$

$$(Gw)_R = 0.$$

By means of (2.30) and (2.123), equations (2.144) can be rewritten to yield

$$Tw + \sum_{k=0}^{2} (-1)^{(k)}[h_k w^{(k)}]^{(k)} + gw' = 0, \tag{2.145}$$

$$(Gw)_R = 0.$$

Using (1.100) and (1.101), equations (2.145) can be changed into

$$-\lambda y + S_1 y + gS_2 y + gNy = 0, \tag{2.146}$$

$$(G_y)_R = 0.$$

The operators in (2.146) are defined as follows

$$\begin{aligned} S_1 y &= h_0 y + [h_2 y'']'', && \text{selfadjoint, positive definite,} \\ S_2 y &= -[h_1^* y']', && \text{selfadjoint, negative definite,} \\ Ny &= y', && \text{nonselfadjoint.} \end{aligned} \tag{2.147}$$

By means of (2.146), the "energy functional"

$$\mathscr{E} = \int_B [-\lambda y^2 + yS_1 y + gyS_2 y + gyNy]\, dB \tag{2.148}$$

can be set up for the solution y. Obviously,

$$\mathscr{E} \equiv 0. \tag{2.149}$$

Because of (2.149), the eigenvalue λ can be obtained from (2.148) in form of Rayleigh's quotient

$$\lambda = \frac{\int_B [yS_1 y + gyS_2 y + gyNy]\, dB}{\int_B y^2\, dB}. \tag{2.150}$$

Due to the fact that operator N appears in the numerator of (2.150), λ may admit real as well as complex values, as already mentioned. This is contrary to (2.39), where only real values are possible since that case is a conservative one. Assuming for example that

$$\lambda = \alpha + i\beta, \tag{2.151}$$

then, the complex solution

$$y = u + iv \tag{2.152}$$

corresponds to the complex λ. Using (2.151) and (2.152) in (2.150),

$$\beta = \frac{\int_B (uNv - vNu)\,dB}{\int_B (u^2 + v^2)\,dB} \tag{2.153}$$

results. For the calculation involved attention has been paid to the fact that operators S_1 and S_2 are selfadjoint. Obviously, $\beta \neq 0$, since N is a nonselfadjoint operator. Hence, complex values for λ can indeed not be ruled out.

Now, equation (2.150) may be written down for

$$\bar{y} = y + \delta y, \qquad \delta y = \alpha\eta, \tag{2.154}$$

where η is an admissible function and α a parameter. Then, the first variation of the eigenvalue is obtained as

$$\delta\lambda = \alpha \left[\frac{\partial\lambda(\bar{y})}{\partial\alpha} \right]_{\alpha=0} \tag{2.155}$$

which yields

$$\delta\lambda = \frac{g \int_B (yN\delta y - \delta yNy)\,dB}{\int_B y^2\,dB} \;. \tag{2.156}$$

The second variation is

$$\delta^2\lambda = \alpha^2 \left[\frac{\partial^2\lambda(\bar{y})}{\partial^2\alpha} \right]_{\alpha=0} \tag{2.157}$$

or

$$\delta^2\lambda = 2\,\frac{\int_B \delta y\,(S_1\delta y + gS_2\delta y + gN\delta y - \lambda\delta y)\,dB}{\int_B y^2\,dB} - 4\delta\lambda\,\frac{\int_B y\delta y\,dB}{\int_B y^2\,dB}\;. \tag{2.158}$$

Due to the nonselfadjoint character of N, the variation $\delta\lambda$ is different from zero according to (2.156). In addition, $\delta^2\lambda$ is indefinite according to (2.158). All this adds up to the conclusion that λ, as calculated by means of (2.150), is neither extremal nor stationary in the nonconservative case. This result had to be expected, since it had already been shown that (2.128) is also not an extremum principle.

The appearance of N causes in the nonconservative case the difference in the behaviour of λ as compared with its behaviour in the conservative case which was described in Section 2.1. Therefore, it is reasonable to consider gNy in (2.146) as a "perturbation term". Hence, let

$$-\lambda y + S_1 y + gS_2 y + \epsilon g N y = 0, \tag{2.159}$$

where ϵ is a "perturbation parameter". Assume that for $\epsilon = 0$, i.e., for the conservative problem, the solution be y_0 and the eigenvalue λ_0. Then, an expansion with respect to ϵ yields

$$\lambda = \lambda_0 + \epsilon\lambda_1 + \epsilon^2\lambda_2 + \ldots, \tag{2.160}$$

$$y = y_0 + \epsilon y_1 + \epsilon^2 y_2 + \ldots. \tag{2.161}$$

Using (2.160) and (2.161) in (2.159), and collecting and equating terms multiplied by the same power of ϵ, yields

$$S_1 y_1 + gS_2 y_1 - \lambda_0 y_1 = \lambda_1 y_0 - gNy_0, \tag{2.162}$$

$$S_1 y_2 + gS_2 y_2 - \lambda_0 y_2 = \lambda_1 y_1 + \lambda_2 y_0 - gNy_1, \tag{2.163}$$

etc.

Here, λ_0 and y_0 can be considered as known quantities satisfying the differential equation

$$S_1 y_0 + gS_2 y_0 - \lambda_0 y_0 = 0 \tag{2.164}$$

of the conservative case.

Multiplying equations (2.162), (2.163) etc. by y_0, and then integrating, the left hand sides of these equations become all equal to zero because of (2.164) and the selfadjointness of S_1, S_2. Therefore,

$$\lambda_1 = g\,\frac{\int\limits_B y_0 N y_0\, dB}{\int\limits_B y_0^2\, dB}, \tag{2.165}$$

$$\lambda_2 = g\left\{\frac{\int\limits_B y_0 N y_1\, dB}{\int\limits_B y_0^2\, dB} - \frac{\left[\int\limits_B y_0 N y_0\, dB\right]\left[\int\limits_B y_0 y_1\, dB\right]}{\left[\int\limits_B y_0^2\, dB\right]^2}\right\} \tag{2.166}$$

etc.

Since (2.159) is a linear differential equation, the expansions (2.160), (2.161) are still valid for $\epsilon = 1$ [22]. Hence, (2.160) yields the eigenvalue of (2.146) for $\epsilon = 1$ if (2.165), (2.166) etc. are used in (2.160). The result is

$$\lambda = \lambda_0 + g\left\{\frac{\int_B y_0 N y_0 \, dB}{\int_B y_0^2 \, dB} + \frac{\int_B y_0 N y_1 \, dB}{\int_B y_0^2 \, dB} - \frac{\left[\int_B y_0 N y_0 \, dB\right]\left[\int_B y_0 y_1 \, dB\right]}{\left[\int_B y_0^2 \, dB\right]^2} + \ldots\right\}. \tag{2.167}$$

Having calculated y_0, y_1, etc. by means of (2.162), (2.164), etc., the eigenvalue λ of the nonconservative problem can be determined according to (2.167) if the eigenvalue λ_0 of the corresponding conservative problem (2.164) is known.

The preceding considerations have been made in order to distinguish a case of a certain interest. That is the case of the *completely supported systems*, i.e., systems which are supported along their whole edge R. According to (2.147$_3$),

$$Ny = \frac{\partial y}{\partial x_i}, \tag{2.168}$$

and due to the support of the edge

$$(y)_R = 0. \tag{2.169}$$

Therefore, using (1.21) and taking (2.168) and (2.169) into account, one finds that

$$\int_B y N y \, dB = 0. \tag{2.170}$$

This relationship holds also true for all comparison functions of y.

For completely supported systems, relationship (2.150) changes immediately into

$$\lambda = \frac{\int_B [y S_1 y + g y S_2 y] \, dB}{\int_B y^2 \, dB}. \tag{2.39}$$

That is an expression which has already been encountered in the context of conservative systems. However, the agreement is only of a formal nature. Because (2.39) has to be set up with the respective true *solution* y, and this solution is obtained in the nonconservative case from (2.146), while in the conservative case it follows from (2.33), (2.36). Hence, the solution to be used in (2.39) is from case to case different. Nevertheless, one is not able to distinguish for completely supported conservative and nonconservative systems between their approximation $\bar{\lambda}$ which are obtained from (2.39) by using comparison functions $\bar{y}$ instead of the solution y. The reason is that in both cases one does have *identical comparison functions* $\bar{y}$ as *the boundary conditions are the same* in both cases.

The identity of approximations $\bar{\lambda}$ in the case of completely supported conservative and nonconservative systems indicates a close relationship between the exact eigenvalues λ of these two cases. This relationship indeed exists. Calling the eigenvalue of the conservative system λ_0 and the eigenvalue of the corresponding nonconservative system λ, one can use (2.160) to check their mutual relationship. For completely supported systems, (2.170) is also valid for $y = y_0$. Therefore, $\lambda_1 \equiv 0$ according to (2.165), and

$$\lambda = \lambda_0 + \epsilon^2 \lambda_2 + \dots . \tag{2.171}$$

Quantities λ and λ_0 are different from each other only by a quantity involving the second order of ϵ. Hence, they must be close together. For $\epsilon = 1$, equation (2.167) yields

$$\lambda = \lambda_0 + g \frac{\int_B y_0 N y_1 \, dB}{\int_B y^2 \, dB} + \dots , \tag{2.172}$$

because (2.170) holds true for y_0.

For a further classification of nonconservative systems consider the energy functional $\mathscr{E}$. Condition

$$\mathscr{E} = 0 \tag{2.149}$$

is satisfied for the solution y only. Writing $\mathscr{E}$ down for a comparison function ϕ yields

$$\mathscr{E}[\phi] = \int_B [-\lambda\phi^2 + \phi S_1 \phi + g\phi S_2 \phi + g\phi N\phi] \, dB, \tag{2.173}$$

and

$$\mathscr{E}[\phi] = 0 \tag{2.174}$$

can be satisfied approximately only. Nothing can be said immediately about magnitude and sign of the error in (2.174) since λ is neither extremal nor stationary. Convergence considerations concerning the approximate calculation of λ will be made later on. However, in anticipation of facts to be stated later on, one is entitled already now to assert that the error will be small, specifically for a linear combination

$$y \sim a\phi + b\psi \tag{2.175}$$

of independent, orthonormal comparison function ϕ and ψ satisfying

$$\int_B \phi^2 \, dB = \int_B \psi^2 \, dB = 1, \qquad \int_B \phi\psi dB = 0. \tag{2.176}$$

Using (2.175) in (2.148) yields

$$\begin{aligned} &\mathscr{E}[a\phi + b\psi] = \\ &= a^2\{\mathscr{E}[\phi] + ab\{\mathscr{E}[\phi, \psi]\} + ba\{\mathscr{E}[\psi, \phi]\} + b^2 \mathscr{E}[\psi], \end{aligned} \tag{2.177}$$

where $\mathscr{E}[\phi]$ and $\mathscr{E}[\psi]$ are defined by (2.173), whereas $\mathscr{E}[\phi, \psi]$ and $\mathscr{E}[\psi, \phi]$ are defined by

$$\mathscr{E}[\phi, \psi] = \int_B [-\lambda\phi\psi + \phi S_1 \psi + g\phi S_2 \psi + g\phi N\psi] dB. \tag{2.178}$$

Moreover,

$$\mathscr{E}[a\phi + b\psi] = 0 \tag{2.179}$$

shall hold true approximately. Since

$$\mathscr{E}[\phi, \psi] \neq \mathscr{E}[\psi, \phi], \tag{2.180}$$

$\mathscr{E}[a\phi + b\psi]$ is not a quadratic but a *bilinear form*. Hence, relationships (2.177), (2.179) must necessarily be representable as

$$[a\{\mathscr{E}[\phi]\} + b\{\mathscr{E}[\phi, \psi]\}]a + [a\{\mathscr{E}[\psi, \phi]\} + b\{\mathscr{E}[\psi]\}]b = 0. \tag{2.181}$$

Because of the arbitrariness of the coefficients a and b, (2.181) can be correct only if

$$\begin{aligned} &a\{\mathscr{E}[\phi]\} + b\{\mathscr{E}[\phi, \psi] = 0 \\ &a\{\mathscr{E}[\psi, \phi]\} + b\{\mathscr{E}[\psi]\} = 0. \end{aligned} \tag{2.182}$$

Since a and b must not be zero at the same time, the determinant of the system must be zero, as the system is homogeneous. Hence,

$$\mathscr{E}[\phi]\,\mathscr{E}[\psi] = \mathscr{E}[\phi, \psi].\,\mathscr{E}[\psi, \phi]. \tag{2.183}$$

Conditions

$$\mathscr{E}[\phi] = 0, \qquad \mathscr{E}[\psi] = 0 \tag{2.184}$$

are straight lines in the (g, λ) plane, namely

$$g = -\frac{C_{1\phi}}{C_{2\phi}} + \frac{\lambda}{C_{2\phi}}, g = -\frac{C_{1\psi}}{C_{2\psi}} + \frac{\lambda}{C_{2\psi}}, \tag{2.185}$$

which are obtained according to (2.173), (2.176) with the coefficients

$$\begin{aligned} C_{1\phi} &= \int_B \phi S_1 \phi \, dB, & C_{2\phi} &= \int_B [\phi S_2 \phi + \phi N \phi] dB, \\ C_{1\psi} &= \int_B \psi S_1 \psi \, dB, & C_{2\psi} &= \int_B [\psi S_2 \psi + \psi N \psi] dB. \end{aligned} \tag{2.186}$$

The straight lines (2.184) subdivide the (g, λ) plane into four domains

$$\begin{aligned} &\text{Domain I, where } \mathscr{E}[\phi] > 0, \mathscr{E}[\psi] > 0, \mathscr{E}[\phi]\mathscr{E}[\psi] > 0, \\ &\text{Domain II, where } \mathscr{E}[\phi] < 0, \mathscr{E}[\psi] < 0, \mathscr{E}[\phi]\mathscr{E}[\psi] > 0, \\ &\text{Domain III, where } \mathscr{E}[\phi] > 0, \mathscr{E}[\psi] < 0, \mathscr{E}[\phi]\mathscr{E}[\psi] < 0, \\ &\text{Domain IV, where } \mathscr{E}[\phi] < 0, \mathscr{E}[\psi] > 0, \mathscr{E}[\phi]\mathscr{E}[\psi] < 0. \end{aligned} \tag{2.187}$$

In which of these domains the trace of the eigenvalue curve $\lambda = \lambda(g)$ determined by (2.150) will be located depends according to (2.187) and (2.183) on the sign of the expression.

$$\mathscr{R} = \mathscr{E}[\phi, \psi].\,\mathscr{E}[\psi, \phi]. \tag{2.188}$$

Selecting without restriction of generality ϕ and ψ as the first two eigenfunctions of the free vibration of the system, then, not only (2.176) holds true, but also

$$\int_B \phi S_1 \psi \, dB = \int_B \psi S_1 \phi \, dB = 0, \tag{2.189}$$

taking definition (2.147) of S_1 into account. Hence,

$$\mathscr{R} = g^2 \left[\int_B \phi (S_2 + N) \psi \, dB \right] \left[\int_B \psi (S_2 + N) \phi \, dB \right], \tag{2.190}$$

according to (2.178), (2.176), and (2.189). Also, (2.183) changes into

$$\mathscr{E}[\phi]\,\mathscr{E}[\psi] = g^2 \mathscr{P}, \tag{2.191}$$

using (2.188), (2.190), and

$$\mathscr{P} = \left[\int_B \phi (S_2 + N) \psi \, dB\right]\left[\int_B \psi (S_2 + N) \phi \, dB\right]. \tag{2.192}$$

Finally, it becomes apparent that the domain in which the eigenvalue curve will be located depends on the sign of parameter $\mathscr{P}$.

Relationship (2.191) is one between λ and g and can therefore be considered to be an approximation of the eigenvalue curve. Hence,

$$\widetilde{F}(\lambda, g) = \mathscr{E}[\phi].\mathscr{E}[\psi] - g^2 \mathscr{P} = 0. \tag{2.193}$$

As a glance on Figure 8b shows, one has to deal with a flutter system if the eigenvalue curve admits of a horizontal tangent. Whether this is possible or not has to be investigated using (2.193).

Obviously,

$$d\widetilde{F} = \frac{\partial \widetilde{F}}{\partial g} dg + \frac{\partial \widetilde{F}}{\partial \lambda} d\lambda = 0.$$

Hence,

$$\frac{dg}{d\lambda} = - \frac{\partial \widetilde{F}/\partial \lambda}{\partial \widetilde{F}/dg}. \tag{2.194}$$

There is a horizontal tangent and flutter is possible if $dg/d\lambda = 0$: According to (2.194), this is the case when the flutter criterion

$$\frac{d\widetilde{F}}{d\lambda} = 0 \tag{2.195}$$

is satisfied. Evaluating (2.195) by means of (2.193) using (2.191), (2.173), (2.176) yields

$$\frac{\partial \widetilde{F}}{\partial \lambda} = -\{\mathscr{E}[\phi] + \mathscr{E}[\psi]\} = 0. \tag{2.196}$$

This is a straight line bisecting the sector which is bordered by the lines (2.184). The line (2.196) is therefore located in domains III and IV. It is the geometric locus for the horizontal tangent of (2.193) which indicates flutter. Therefore it will be called *flutter line*. One can now conclude that flutter is only possible for those eigenvalue curves which are located in domains III and IV, thus intersecting the flutter line.

Consider again completely supported systems, for example those shown in Figure 10a, b. For these systems, especially

$$C_{2\phi} = \int_B \phi S_2 \phi \, dB < 0, \qquad C_{2\psi} = \int_B \psi S_2 \psi \, dB < 0, \tag{2.197}$$

since condition (2.170) holds true for ϕ and ψ, and S_2 is a negative definite operator. Therefore, the straight lines (2.184) and (2.196) have a negative slope, so that the conditions in Figure 11 prevail.

If $\mathscr{P} > 0$ for the completely supported systems, as shown in Figure 12, the trace of their eigenvalue curves is completely located in domains I and II. Hence, these eigenvalue curves can never intersect the flutter line (2.196), which is located in domains III and IV. Thus, flutter is excluded. However, the eigenvalue curves cross the g-axis yielding the buckling loads $g_{1\,\mathrm{crit}}, g_{2\,\mathrm{crit}}$. These are by their very nature divergence loads. The conclusion is that *completely supported systems for which $\mathscr{P} > 0$ are divergence type systems.*

If $\mathscr{P} < 0$ for completely supported systems, the trace of their eigenvalue curves is in domains III and IV. The flutter line is being intersected and flutter occurs. Due to the complete support, the lines (2.184) limiting the domains are sloped in such a way that the eigenvalue curves are possibly able to intersect the g-axis at $g > 0$. Let

$$\mathscr{S} = \frac{(C_{1\phi}C_{2\psi} - C_{2\phi}C_{1\psi})^2}{4C_{1\phi}C_{1\psi}}, \tag{2.200}$$

than the g-axis will be intersected if

$$\mathscr{P} > -\mathscr{S}. \tag{2.201}$$

Consequently, instability by divergence occurs in addition to flutter. As Figure 13 shows, the divergence buckling loads $g_{1\,\mathrm{crit}}, g_{2\,\mathrm{crit}}$ are smaller than the flutter load g_{f1} so that a system of that kind behaves pracically like a divergence type system. It may therefore be called a *pseudo-divergent* system.

If $\mathscr{P} < -\mathscr{S}$ were possible for completely supported systems, the g-axis would not be intersected, divergence buckling loads were ruled out, and the system concerned were a *pure flutter system* as shown in Figure 14.

For not completely supported systems, using (2.168) and (1.21) in their

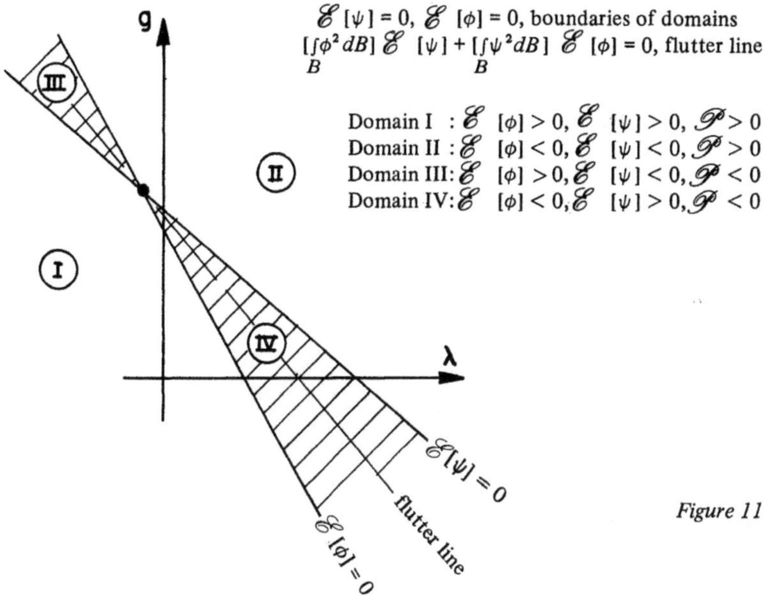

Figure 11

Figure 10a

Figure 10b

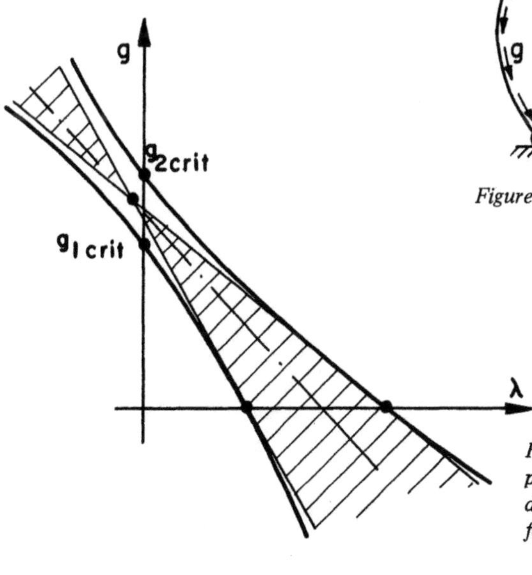

Figure 12. Completely supported, nonconservative divergence type system for $\mathscr{P} > 0$

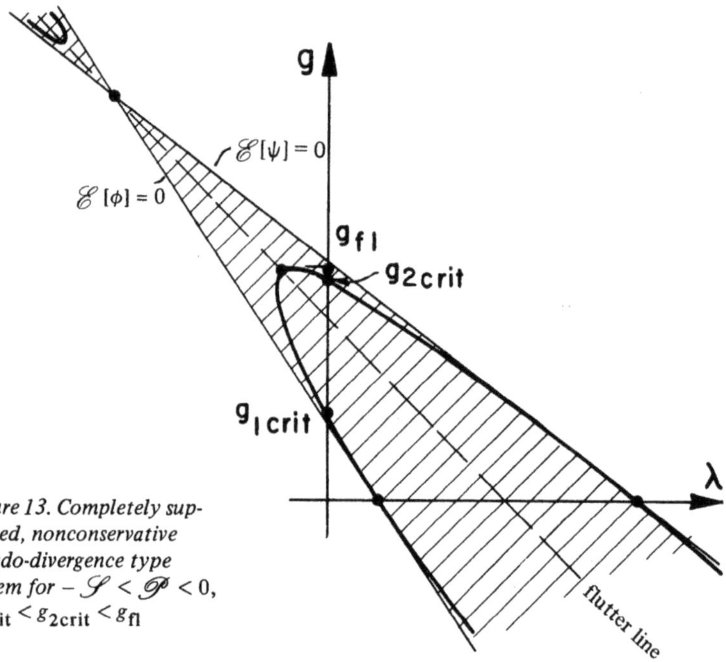

Figure 13. Completely supported, nonconservative pseudo-divergence type system for $-\mathscr{S} < \mathscr{P} < 0$, $g_{1crit} < g_{2crit} < g_{fl}$

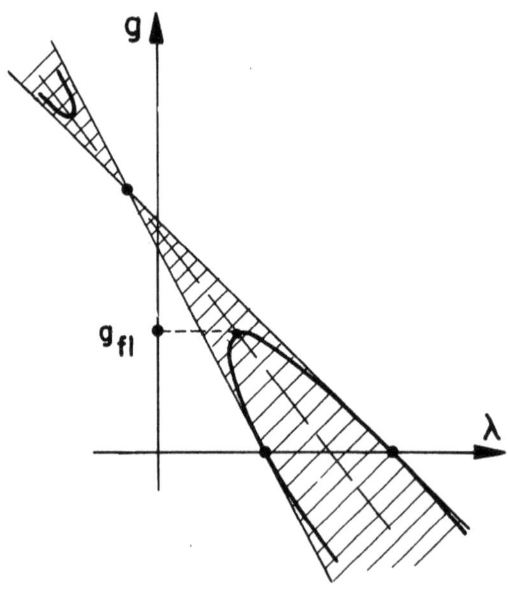

Figure 14. Completely supported, nonconservative flutter type system for $\mathscr{P} < -\mathscr{S}$

one-dimensional version, one finds that

$$\int_B y\,Ny\,dB = \frac{1}{2}\,y^2\,(l) > 0, \tag{2.202}$$

where l is the length of the system. In the derivation of (2.202) it has in addition been assumed that at least $y\,(0) = 0$, as the system is not supposed to float freely in the space.

The further behaviour of the system is guided by the functional (2.202). For example, let

$$\int_B \phi N \phi\, dB < - \int_B \phi S_2 \phi\, dB, \tag{2.203}$$

then, $C_{2\phi} < 0$. In addition, assume $C_{2\psi} < 0$. Then, conditions like for completely supported systems would prevail: for $\mathscr{P} > 0$ there were divergence, for $-\mathscr{S} < \mathscr{P} < 0$ pseudo-divergence, and for $\mathscr{P} < -\mathscr{S}$ flutter.

If, however,

$$\int_B \phi N \phi\, dB < - \int_B \phi S_2 \phi\, dB, \tag{2.204}$$

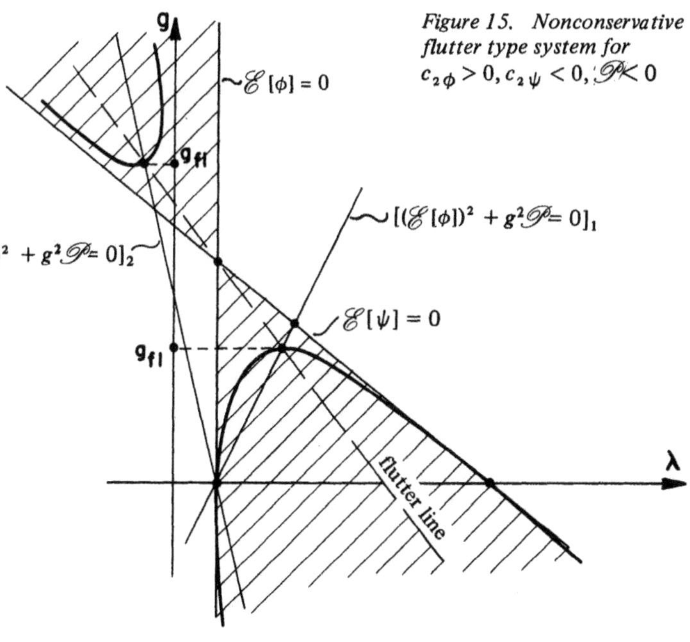

Figure 15. Nonconservative flutter type system for $c_{2\phi} > 0, c_{2\psi} < 0, \mathscr{P} < 0$

then, $C_{2\phi} > 0$. In addition, assume $C_{2\psi} < 0$. Then, line $\mathscr{E}[\phi] = 0$ has a positive slope, and line $\mathscr{E}[\psi] = 0$ has a negative one. The possibility of $\mathscr{P} > 0$ be excluded for not completely supported systems. This assumption is reasonable as will be shown by means of examples in Section 2.3. However, for $\mathscr{P} < 0$ the trace of the eigenvalue curve is located in domains III and IV as shown in Figure 15. Since the flutter line is intersected, horizontal tangents do exist, and the smallest positive buckling load is a flutter load. That means, one is dealing with a flutter system.

In order to determine the flutter load, relationships (2.192) and (2.196) will be used simultaneously. The first equation states that the flutter load g_{f1} is an ordinate of the eigenvalue curve, and the second equation states that it is specifically the ordinate of that point of the eigenvalue curve which has a horizontal tangent. Conditions (2.192), (2.196) together yield for example

$$\{\mathscr{E}[\phi]\}^2 + g^2 \mathscr{P} = 0, \tag{2.205}$$

which is a pair of straight lines that intersect together with the flutter line the eigenvalue curve in a point having a horizontal tangent. By means of the flutter line (2.196) and the pair of lines (2.205), the flutter load g_{f1} can be determined without the whole eigenvalue curve having to be drawn.

For the sake of completeness, let the cases $C_{2\phi} > 0$, $C_{2\psi} > 0$ (Figure 16) and $C_{2\phi} < 0$, $C_{2\psi} < 0$ (Figure 17) be mentioned. As can be seen in the figures, these cases lead to negative divergence and flutter loads. In Section 2.3 it will become apparent that these cases are irrelevant as they do not occur for beams. Therefore, they may be disregarded in the following.

Summarizing it may be stated:
completely supported nonconservative systems behave very closely like conservative systems. For $\mathscr{P} > 0$ they are divergence type systems like the conservative ones. For $-\mathscr{S} < \mathscr{P} < 0$ they are at least pseudo-divergence type systems which practically behave like divergence type systems. If, however, the support is not complete and $C_{2\phi} > 0$, $C_{2\psi} < 0$, the whole difference to conservative systems comes to light. Then, only $\mathscr{P} < 0$ is of practical importance, and one has to deal with a flutter system, as examples will show. The flutter load is obtained by intersecting the flutter line (2.196) with the pair of lines (2.205).

It might be of interest to notice that there are nonconservative problems for which there exists a *generalized Rayleigh quotient* and which are strictly of the divergence type. For problems of this kind, the operators S_1 and $S_2 + N$ are supposed to be positive definite, negative definite, respectively, and selfadjoint

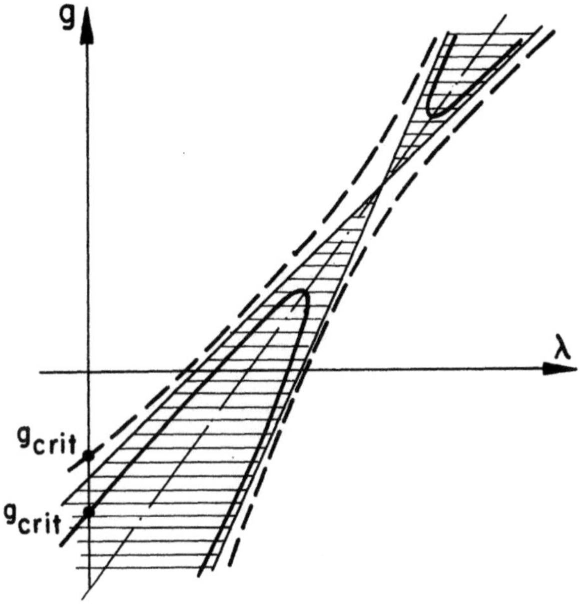

Figure 16

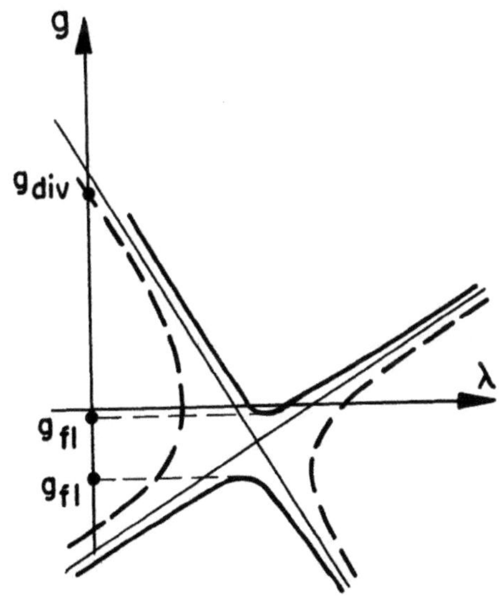

Figure 17

with respect to the boundary conditions of the problem. Moreover, for any pair of comparison functions, u, v, the following is assumed to hold:

$$\int_B (S_1 u)\,[(S_2 + Nv]\,dB = \int_B (S_1 v)\,[(S_2 + N)u]\,dB,$$

$$\int_B (S_1 u)\,[(S_2 + N)u]\,dB < 0.$$

These two conditions result in the operator S_1 to be *generalized selfadjoint* and *generalized negative definite* with respect to operator $S_2 + N$.

For the sake of simplicity let $S_2 + N \equiv T$ be used in the following.

Now, assume the eigenvalue problem

$$-\lambda y + S_1 y + g(S_2 + N)y \equiv -\lambda y + S_1 y + gTy = 0$$

to be given. Consider the functional

$$\int_B (-\lambda y + S_1 y + gTy)\,Ty\,dB = 0.$$

Varying this functional, and taking the selfadjointness of $S_2 + N \equiv T$ as well as the generalized selfadjointness of S_1 with respect to T into account, yields after a simple calculation

$$-d\lambda \int_B yTy\,dB + dg \int_B (Ty)^2\,dB = 0.$$

Hence, for the slope of the eigenvalue curve in the (g, λ) plane follows

$$\frac{dg}{d\lambda} = \frac{\int_B yTy\,dB}{\int_B (Ty)^2\,dB} < 0.$$

Since the slope is always negative, a horizontal tangent for the eigenvalue curve and thus flutter for the system are ruled out. Therefore, the system must be a divergence type system.

By virtue of this property of the system, the trace of the eigenvalue curve is as in Figure 8a. That means, the buckling load g_{crit} occurs at $\lambda = 0$. But then, it is sufficient to work with

$$S_1 y + gTy = 0$$

for the evaluation of g_{crit}.

Now, the generalized Rayleigh quotient R_g may be introduced. It is obtained by solving the identity

$$\int_B [(S_1 y)(Ty) + g(Ty)^2]dB = 0$$

for g, i.e., by establishing

$$g = -\frac{\int_B (S_1 y)(Ty)\,dB}{\int_B (Ty)^2\,dB},$$

and substituting therein the *comparison function* z for y. The result is

$$R_g = -\frac{\int_B (S_1 z)(Tz)\,dB}{\int_B (Tz)^2\,dB}.$$

R_g is always positive, as operator S_1 is assumed to be negative definite with respect to operator T.

Let R_g be varied by varying z. This operation results in

$$\delta R_g = \frac{D\int_B [-(S_1 z)(T\delta z) - (S_1 \delta z)(Tz)]dB - N\int_B 2(Tz)(T\delta z)dB}{D^2}$$

where N is the numerator and D the denominator of R_g. Using the generalized selfadjointness of S_1 with respect to T, and the fact that $N/D = R_g$, yields

$$D \cdot \delta R_g = -2\int_B [(S_1 z + R_g Tz)(T\delta z)]dB.$$

However, $S_1 y + gTy = 0$, $R_g \to g$ for $z \to y$, and $D \neq 0$. Therefore,

$$\delta R_g = 0 \text{ for } z = y,$$

i.e., R_g *is stationary* at $z = y$, and can be used as a generalized Rayleigh quotient.

Pflüger's rod, i.e., the pinned-pinned rod with uniformly distributed follower forces, is an example to this theory. Here $S_1 y = \alpha y^{IV}$, $(S_2 + N)y \equiv Ty = y''(l-x)$, and the boundary conditions are $y(0) = y(l) = y''(0) = y''(l) = 0$. It can easily be checked that S_1 and T do indeed possess all required properties. Using the comparison function $z = \sin \pi x/l$ in R_g yields

$$R_g = 19.74\alpha/l^3 > 18.96\alpha/l^3 = g_{1\,\text{crit}}.$$

The value of $g_{1\,\mathrm{crit}} = 18.96\alpha/l^3$ is well known in the literature [23]. Thus, the extremum property of R_g is verified. Further information on the generalized Rayleigh quotient and on so called *conservative systems of the second kind* can be found in the literature [24].

All the previous considerations and the discussions of the possible trace of eigenvalue curves are based on the assumption that the equations used for the calculation of the eigenvalues admit of a convergent solution. Therefore, the convergence of the system (2.132) of the equations of Ritz must be investigated in order to justify the assumption. These equations written in full read

$$\sum_i a_i\{-\lambda\delta_{ij} + \int_B [h_0\phi_i\phi_j + h_1\phi_i'\phi_j' + h_2\phi_i''\phi_j'' + g\phi_i'\phi_j]dB + \int_{R_F} P\phi_i'\phi_j\,dR\} = 0. \quad (2.206)$$

Using

$$\int_B \phi_i\phi_j\,dB = \delta_{ij} \quad (1.112)$$

and integrating (2.206) by parts yields

$$\sum_i a_i\{\int_B [(-\lambda + h_0)\phi_i - (h_1\phi_i')' + (h_2\phi_i'')'' + g\phi_i']\phi_j dB + [h_1\phi_i'\phi_j + h_2\phi_i''\phi_j' - (h_2\phi_i'')'\phi_j]_R + \int_{R_F} P\phi_i'\phi_j\,dR\} = 0. \quad (2.207)$$

For one dimensional systems,

$$B \text{ is the interval } 0 \leqq x \leqq l; \quad (2.208)$$

furthermore,

$$R \text{ consists of the boundary points } x = 0,\ x = l, \quad (2.209)$$

and for an arbitrary function f

$$\int_{R_F} f dR \text{ degenerates into the boundary value } f(l). \quad (2.210)$$

For the following let

$$h_1 = -\phi(x) - F, \qquad \phi(x) \geqq 0,\ \phi(l) = 0,\ F = \text{const}, \quad (2.211)$$

so that

$$h_1(l) = -F. \tag{2.212}$$

In addition let

$$h_0 = 0, h_2 = \alpha = \text{const.}, P = \gamma F, 0 < \gamma < 1. \tag{2.213}$$

Then, equation (2.207) changes into

$$\sum_i a_i \{ \int_0^l [-\lambda\phi_i - (h_1\phi_i')' + h_2\phi_i^{IV} + g\phi_i']\phi_j dx$$

$$+ [h_1\phi_i'\phi_j + h_2\phi_i''\phi_j' - h_2\phi_i'''\phi_j]_R + \gamma F\phi_i'(l)\phi_j(l)\} = 0. \tag{2.214}$$

The boundary conditions of the posed problem be for example

$$y(0) = y'(0) = y''(l) = y'''(l) + h_2^{-1}(1-\gamma)Fy'(l) = 0. \tag{2.215}$$

For the calculation, the coordinate functions ϕ_i in (2.214) may be chosen such that they are only *admissible* with respect to (2.215), i.e., they may not satisfy all of the dynamic boundary conditions or even none of them. For example, let the auxiliary problem

$$\phi_i^{IV} = \omega_i\phi_i, \tag{2.216}$$

$$\phi_i(0) = \phi_i'(0) = \phi_i''(l) = \phi_i'''(l) = 0 \tag{2.217}$$

be introduced. Its eigenfunctions ϕ_i are selected as coordinate functions. A comparison of (2.217) with (2.215) shows that the ϕ_i violate indeed the last dynamic boundary condition of the posed problem. Therefore, and because of (2.211), (2.209) equation (2.214) changes into

$$\sum_i a_i \{ \int_0^l [-\lambda\phi_i - (h_1\phi_i')' + h_2\phi_i^{IV} + g\phi_i']\phi_j dx + (\gamma-1)F\phi_i'(l)\phi_j(l)\} = 0. \tag{2.218}$$

In the view of mechanics, (2.216), (2.217) represent the free vibrations of the unloaded given system. Since problem (2.216), (2.217) is selfadjoint and positive definite, a symmetric Green function exists,

$$G(x, \xi) = \sum_k \frac{\phi_k(x)\phi_k(\xi)}{\omega_k}, \tag{2.219}$$

which is bilinearly expandable in terms of the eigenfunctions ϕ_i of (2.216), (2.217). In addition, the ϕ_i are a complete system of functions with respect to the solution y of the posed problem. This will be proved as follows:

Assume

$$y(x) = \sum_i a_i \phi_i(x). \tag{2.220}$$

Since $y(x)$ is an admissible function for the auxiliary problem, as it satisfies the geometric boundary conditions of (2.217), also

$$r_p(x) = y(x) - \sum_{i=1}^{p} a_i \phi_i(x) \tag{2.221}$$

is only an admissible function. It can be used to calculate Rayleigh's quotient $R[r_p(x)]$ of problem (2.216), (2.217), i.e.,

$$R[r_p(x)] = \frac{\int_0^l r_p''^2\, dx}{\int_0^l r_p^2\, dx} \geqq \omega_1. \tag{2.222}$$

Since according to (2.221)

$$\int_0^l r_p \phi_k\, dx = \int_0^l y\phi_k\, dx - \sum_{i=1}^{p} a_i \int_0^l \phi_i \phi_k\, dx, \tag{2.223}$$

and since because of (2.220)

$$\int_0^l y\phi_k\, dx = \sum_{i=1}^{p} a_i \int_0^l \phi_i \phi_k\, dx, \tag{2.224}$$

(2.223) yields

$$\int_0^l r_p \phi_k\, dx = a_k - a_k = 0, \quad k = 1,2,3,\ldots,p, \tag{2.225}$$

using also (1.112). Hence, r_p is orthogonal to all eigenfunctions ϕ_k, $k = 1,2,3, \ldots, p$. But then, not only (2.222) but even

$$R[r_p(x)] = \frac{\int_0^l r_p''^2\, dx}{\int_0^l r_p^2\, dx} \geqq \omega_{p+1} \tag{2.226}$$

holds true, consequently,

$$\int_0^l r_p^{\,2}\,dx \leqq \frac{\int_0^l r_p''^2\,dx}{\omega_{p+1}}\,. \tag{2.227}$$

By virtue of (2.221),

$$\int_0^l y''^2\,dx = \int_0^l \left(r_p'' + \sum_{i=1}^p a_i\phi_i''\right)^2 dx$$

$$= \int_0^l r_p''^2\,dx + 2\sum_{i=1}^p a_i \int_0^l r_p''\phi_i''\,dx + \sum_{i=1}^p a_i^2 \int_0^l [\phi_i'']^2\,dx. \tag{2.228}$$

But,

$$\int_0^l r_p''\phi_i''\,dx = \Big[r_p'\phi_i'' - r_p\phi_i'''\Big]_0^l + \omega_i \int_0^l r_p\phi_i\,dx = 0, \tag{2.229}$$

according to $r_p'(0) = r_p(0) = 0$, (2.216), (2.217) and (2.225). Moreover,

$$\int_0^l \phi_i''^2\,dx = \Big[\phi_i'\phi_i'' - \phi_i\phi_i'''\Big]_0^l + \int_0^l \phi_i\phi_i^{\mathrm{IV}}\,dx = \omega_i, \tag{2.230}$$

which follows from (2.217), (2.216), and (1.112).

Using (2.229) and (2.230) in (2.228),

$$\int_0^l y''^2\,dx = \int_0^l r_p''^2\,dx + \sum_{i=1}^p a_i^2\,\omega_i \tag{2.231}$$

remains. Under the assumption that y'' is square integrable, (2.231) yields

$$\int_0^l r_p''^2\,dx < K, \text{ bounded,} \tag{2.232}$$

and

$$\sum_{i=1}^\infty a_i^2\,\omega_i \leqq y''^2\,dx, \text{ bounded,} \tag{2.233}$$

i.e., the series

$$\sum_i a_i^2 \, \omega_i$$

is *convergent*.

By means of (2.232), inequality (2.227) yields

$$\int_0^l r_p^2 \, dx < \frac{K}{\omega_{p+1}} \tag{2.234}$$

so that

$$\lim \int_0^l r_p^2 \, dx = 0 \text{ for } p \to \infty, \tag{2.235}$$

because the eigenvalues ω_i of (2.216), (2.217) tend toward infinity for increasing subscript i.

Now let (2.221) be rewritten to yield

$$\int_0^l y^2 \, dx = \int_0^l \left(r_p + \sum_{i=1}^{p} a_i \phi_i \right)^2 dx$$

$$= \int_0^l r_p^2 \, dx + 2 \sum_{i=1}^{p} a_i \int_0^l r_p \phi_i \, dx + \int_{i=1}^{p} a_i^2 \int_0^l \phi_i \, dx. \tag{2.236}$$

Because of (1.112) and (2.225) this is equivalent to

$$\int_0^l y^2 \, dx = \int_0^l r_p^2 \, dx + \sum_{i=1}^{p} a_i^2 , \tag{2.237}$$

and because of (2.235), *Parseval's formula*

$$\sum_{i=1}^{\infty} a_i^2 = \int_0^l y^2 \, dx \tag{2.238}$$

is obtained, which proves that the system of functions ϕ_i is complete.

Take a second arbitrary admissible function η and its series expansion

$$\eta \sim \sum_i b_i \phi_i \tag{2.239}$$

into consideration. Then, in correspondence to (2.238), also

$$\sum_{i=1}^{\infty} b_i^2 = \int_0^l \eta^2\,dx. \tag{2.240}$$

Since $y + \eta$ is an admissible function too, Parseval's formula applies to this sum and yields

$$\int_0^l (y+\eta)^2\,dx = \int_0^l y^2\,dx + 2\int_0^l y\eta\,dx + \int_0^l \eta^2\,dx = \sum_{i=1}^{\infty} (a_i + b_i)^2. \tag{2.241}$$

By virtue of (2.238), (2.240),

$$\int_0^l y\eta\,dx = \sum_{i=1}^{\infty} a_i b_i \tag{2.242}$$

follows from (2.241). L. Collatz [25] has shown that for the selfadjoint, positive definite, auxiliary problem (2.116), (2.117) and its Green function $G(x, \xi)$ the inequalities

$$\left(\frac{\partial^{2\nu} G(x,\xi)}{\partial x^{\nu}\,\partial \xi^{\nu}}\right)_{x=\xi} \geqq \sum_{i=1}^{p} \frac{[\phi_i^{(\nu)}(x)]^2}{\omega_i}, \quad \nu = 0,1 \tag{2.243}$$

hold true. This fact will be used to evaluate the convergence of

$$\sum_i a_i \phi_i^{(\nu)}, \quad \nu = 0,1.$$

According to the inequality of Schwarz,

$$\left(\sum_{i=p}^{q} |a_i \phi_i^{(\nu)}(x)|\right)^2 = \left(\sum_{i=p}^{q} |a_i| \sqrt{\omega_i}\,\frac{|\phi_i^{(\nu)}(x)|}{\sqrt{\omega_i}}\right)^2 \leqq \left(\sum_{i=p}^{q} a_i^2\,\omega_i\right)\left(\sum_{i=p}^{q} \frac{[\phi_i^{(\nu)}(x)]^2}{\omega_i}\right).$$

The first factor on the right hand side tends toward zero for $p, q \to \infty$ by virtue of (2.233), and the second factor remains bounded, uniformly in x, because of (2.243). Hence,

$$\sum_{i=p}^{q} |a_i \phi_i^{(\nu)}(x)| \to 0 \text{ for } p, q \to \infty, \quad \nu = 0,1.$$

Consequently, the series

$$\sum_i a_i \phi_i^{(\nu)}, \quad \nu = 0,1, \tag{2.244}$$

are absolutely and uniformly convergent

$$\text{in } 0 \leqq x \leqq l.$$

Now, return to (2.242). By definition,

$$b_i = \int_0^l \eta\phi_i\,dx. \tag{2.245}$$

Using this relationship in (2.242) yields

$$\int_0^l y\eta\,dx = \sum_i a_i \int_0^l \eta\phi_i\,dx. \tag{2.246}$$

Due to the convergence of (2.244) for $\nu = 0$, summation and integration are interchangeable in (2.246) on the right hand side. Thus, (2.246) can be rewritten to yield

$$\int_0^l (y - \sum_i a_i\phi_i)\,\eta\,dx = 0.$$

This shall hold true for any arbitrary, admissible function η. That is only possible if

$$y = \sum_{i=1}^{\infty} a_i\phi_i, \tag{2.247}$$

i.e., series (2.244) converges for $\nu = 0$ even toward the admissible function y, and that in an absolute and uniform manner as has already been shown. According to [26],

$$y' = \sum_{i=1}^{\infty} a_i\phi_i' \tag{2.248}$$

as a consequence of the absolute and uniform convergence of (2.244) for $\nu = 1$. In the proof of convergence of the method of Ritz, which is to follow, the absolutely and uniformly convergent series (2.247) and (2.248) will prove to be useful.

Because of (2.216) and (1.112),

$$\sum_i a_i \int_0^l \phi_i^{\mathrm{IV}}\phi_j\,dx = a_j\omega_j. \tag{2.249}$$

In addition, let

$$\sum_i a_i \int_0^l [-\lambda\phi_i - (h_1\phi_i')' + g\phi_i']\phi_j\,dx = -\int_0^l \Phi(\sum_i a_i\phi_i)\,\phi_j\,dx \tag{2.250}$$

for the sake of brevity. Using (2.249) and (2.250) in (2.218) yields after rearranging

$$h_2 a_j\omega_j = \int_0^l \Phi(\sum_i a_i\phi_i)\phi_j\,dx - (\gamma-1)F\left(\sum_i a_i\,\phi_i'(l)\right)\phi_j(l), \tag{2.251}$$

$$a_j = \int_0^l \frac{1}{h_2}\,\Phi(\sum_i a_i\phi_i)\,\frac{\phi_j}{\omega_j}\,dx - (\gamma-1)\,\frac{F}{h_2}\left(\sum_i a_i\phi_i'(l)\right)\frac{\phi_j(l)}{\omega_j}\,. \tag{2.252}$$

Multiplying both sides of (2.252) by $\phi_j(\xi)$ and adding over j yields

$$\sum_j a_j\phi_j(\xi) \tag{2.253}$$

$$= \int_0^l \frac{1}{h_2}\,\Phi\left(\sum_i a_i\phi_i(x)\right)\sum_j \frac{\phi_j(x)\phi_j(\xi)}{\omega_j}\,dx - (\gamma-1)\frac{F}{h_2}\left(\sum_i a_i\phi_i'(l)\right)\sum_j \frac{\phi_j(l)\phi_j(\xi)}{\omega_j}\,.$$

However,

$$y_n(\xi) = \sum_{j=1}^n a_j\phi_j(\xi) \tag{2.254}$$

is an approximation of the solution y of the posed problem, which converges toward y according to (2.247). Also,

$$G_n(x,\xi) = \sum_{j=1}^n \frac{\phi_j(x)\phi_j(\xi)}{\omega_j} \tag{2.255}$$

is according to (2.219) an approximation of Green's function $G(x, \xi)$, and by definition the approximation converges toward G. Using (2.254) and (2.255) in (2.253) yields

$$y_n(\xi) = \int_0^l \frac{1}{h_2}\,\Phi\left(y_n(x)\right)G_n(x,\xi)\,dx - (\gamma-1)\,\frac{F}{h_2}\,y_n'(l)G_n(l,\xi). \tag{2.256}$$

The given problem in y can be rewritten by means of

$$y^{\mathrm{IV}} = Ly \tag{2.257}$$

in the form

$$L(y) = \frac{1}{h_2}\,\Phi(y) = [\lambda y + (h_1 y')' - gy']\,\frac{1}{h_2}\,, \tag{2.258}$$

$$y(0) = 0,\, y'(0) = 0,\, y''(l) = 0,\, y'''(l) = \frac{1}{h_2}\,(\gamma - 1)Fy'(l). \tag{2.259}$$

Now, let Green's identity

$$\int_0^l [uL(v) - vL(u)]\,dx = \Big[uv''' - u'v'' - vu''' + v'u''\Big]_0^l \tag{2.260}$$

be applied to

$$u = G(x, \xi), \quad v = y(x), \tag{2.261}$$

where $G(x, \xi)$ satisfies boundary conditions (2.217) and y boundary conditions (2.259). Moreover,

$$L(u) = L[G(x, \xi)] = \delta(x, \xi), \tag{2.262}$$

where $\delta(x, \xi)$ is Dirac's function. Relationship (2.262) holds true because by assumption $G(x, \xi)$ is the Green function of operator L with respect to boundary conditions (2.217).

Using (2.261), (2.262), (2.258) in (2.260) and taking the boundary conditions of y and G into account yields

$$\int_0^l \frac{1}{h_2}\,\Phi(y)\,G(x, \xi)\,dx - \int_0^l y(x)\,\delta(x, \xi)\,dx$$

$$= \left[Gy''' - \frac{\partial G}{\partial x}\,y'' - y\,\frac{\partial^3 G}{\partial x^3} + y'\,\frac{\partial^2 G}{\partial x^2}\right]_0^l$$

and after rearranging

$$y(\xi) = \int_0^l \frac{1}{h_2}\,\Phi\big(y(x)\big)G(x, \xi)\,dx - (\gamma - 1)\,\frac{F}{h_2}\,y'(l)\,G(l, \xi). \tag{2.263}$$

This is an integro-differential equation for the solution y. As a comparison with (2.256) shows, the method of Ritz, which led to (2.256), consists in calculating an approximation y_n of y by using an approximation of the integro-differential equation (2.263). This approximation (2.256) does not differ from that one yielded by the so called method of moments [25], as will be shown in Section 4.

Being derived on the basis of (2.216), (2.217), $G(x, \xi)$ is necessarily continuous. Since $G_n(x, \xi)$ is obtained by a finite bilinear expansion involving continuous coordinate functions ϕ_i, $G(x, \xi)$ is likewise continuous. The continuity implies boundness. Hence,

$$|G(x, \xi)|, |G_n(x, \xi)| \text{ are bounded.} \tag{2.264}$$

Since y follows from (2.263) and y_n from (2.256), these two quantities are continuous as well by virtue of $|G(x, \xi)|$, $|G_n(x, \xi)|$ being continuous and because of the assumption that

$$|h|, |h_1'| \text{ are bounded.} \tag{2.265}$$

Thus, in addition,

$$|y|, |y_n| \text{ are bounded.} \tag{2.266}$$

By means of

$$y(x) = \int_0^x (x-\eta) y''(\eta)\, d\eta,\; y'(x) = \int_0^x y''(\eta)\, d\eta, \tag{2.267}$$

equation (2.263) can be changed into an integral equation for y''. Taking (2.258) into account, one obtains

$$h_2 y''(\xi) = \int_0^l \left[\lambda \int_0^x (x-\eta) y''(\eta)\, d\eta + h_1' \int_0^x y''(\eta)\, d\eta + h_1 y''(x) \right.$$

$$\left. -g \int_0^x y''(\eta)\, d\eta \right] \frac{\partial^2 G(x, \xi)}{\partial \xi^2}\, dx - (\gamma-1) F \left[\frac{\partial^2 G(l, \xi)}{\partial \xi^2}\right] \int_0^l y''(\eta)\, d\eta. \tag{2.268}$$

For $y_n''(\xi)$ there exists a corresponding integral equation in terms of $\partial^2 G_n/\partial \xi^2$.

Due to its connection with (2.216), (2.217), $\partial^2 G(x, \xi)/\partial \xi^2$ is continuous [26]. Since the eigenfunctions generated by (2.216), (2.217) have continuous derivatives ϕ_i'', $\partial^2 G_n(x, \xi)/\partial \xi^2$ is also continuous. Continuity implies boundedness. Hence,

$$\left|\frac{\partial^2 G(x, \xi)}{\partial \xi^2}\right|, \left|\frac{\partial^2 G_n(x, \xi)}{\partial \xi^2}\right| \text{ are bounded.}$$

Quantity y'' follows from (2.268), and quantity y_n'' follows from a corresponding equation whose kernel involves $\partial^2 G_n(x, \xi)/\partial\xi^2$. Therefore, these two quantities are continuous and bounded by virtue of the continuity and boundedness of $\partial^2 G(x, \xi)\, \partial\xi^2$ and $\partial^2 G_n(x, \xi)/\partial\xi^2$. Hence,

$$|y''|, |y_n''| \text{ are bounded.} \tag{2.269}$$

However,

$$y'(x) = \int_0^x y''(x)\,dx, \; y_n'(x) = \int_0^x y_n''\,dx.$$

Thus,

$$|y'| \leqq l\,|y''|_{\max}, \; |y_n'| \leqq l\,|y_n''|_{\max},$$

and

$$|y'|, |y_n'| \text{ are bounded.} \tag{2.270}$$

The following considerations are based on assumptions (2.264), (2.265), (2.266), (2.269), (2.270) and the convergence of G_n, y_n, y_n'.

Introducing the operators

$$T_1 y(\xi) = \int_0^l y(x)\, G(x, \xi)\,dx, \tag{2.271}$$

$$T_2 y(\xi) = \int_0^l [(h_1 y'(x))' - g y'(x)]\, G(x, \xi)\,dx - (\gamma - 1) F y'(l) G(l, \xi), \tag{2.272}$$

equation (2.263) can be rewritten as

$$h_2 y(\xi) = \lambda T_1 y(\xi) + T_2 y(\xi) \tag{2.273}$$

taking in addition (2.258) into account. Correspondingly, the approximative equation (2.256) changes into

$$h_2 y_n = \lambda_n T_{1n} y_n(\xi) + T_{2n} y_n(\xi), \tag{2.274}$$

where

$$T_{1n} y_n(\xi) = \int_0^l y_n(x)\, G_n(x, \xi)\,dx, \tag{2.275}$$

and

$$T_{2n}y_n(\xi) = \int_0^l [(h_1 y_n'(x))' - g y_n'(x)] G_n(x, \xi)\, dx - (\gamma - 1) F y_n'(l) G_n(l, \xi). \tag{2.276}$$

The convergence problem of the method of Ritz may now be replaced by the convergence problem of the operator equation (2.274).

In order to investigate the convergence of λ_n toward λ in this sense,

$$\lambda = \frac{h_2 \int_0^l y^2 d\xi - \int_0^l y T_2 y d\xi}{\int_0^l y T_1 y d\xi} \tag{2.277}$$

is derived from (2.273) and

$$\lambda_n = \frac{h_2 \int_0^l y_n^2\, d\xi - \int_0^l y_n T_{2n} y_n d\xi}{\int_0^l y_n T_{1n} y_n d\xi} \tag{2.278}$$

from (2.274). Introducing for the sake of brevity the functionals

$$\begin{aligned} A_1 &= \int_0^l y T_1 y d\xi, & A_2 &= h_2 \int_0^l y^2 d\xi - \int_0^l y T_2 y d\xi, \\ A_{1n} &= \int_0^l y_n T_{1n} y_n d\xi, & A_{2n} &= h_2 \int_0^l y_n^2 d\xi - \int_0^l y_n T_{2n} y_n d\xi, \end{aligned} \tag{2.279}$$

and using (2.279) in (2.277), (2.278) yields

$$|\lambda - \lambda_n| = \left| \frac{A_2}{A_1} - \frac{A_{2n}}{A_{1n}} \right| = \left| \frac{A_2 A_{1n} - A_1 A_{2n}}{A_1 A_{1n}} \right|.$$

This relationship can easily be changed into

$$|\lambda - \lambda_n| \leqq \frac{|A_{1n} - A_1| \cdot |A_2| + |A_1| \cdot |A_2 - A_{2n}|}{|A_1| \cdot |A_{1n}|}. \tag{2.280}$$

From (2.280) follows that λ_n converges toward λ if

$$|A_1|, |A_{1n}|, |A_2| \text{ are bounded,} \tag{2.281}$$

and

$$|A_{1n}-A_1| \leqq \epsilon \text{ for } n>N, \tag{2.282}$$

$$|A_2-A_{2n}| \leqq \epsilon \text{ for } n>N. \tag{2.283}$$

1)

$$|A_1| = \left| \int_0^l y(\xi) \int_0^l y(x) G(x, \xi)\, dx\, d\xi \right| \leqq l^2 \, |y|^2_{\max} \, |G(x, \xi)|_{\max}.$$

Hence, $|A_1|$ is bounded by virtue of (2.264), (2.266).

2) Accordingly,

$$|A_{1n}| \leqq l^2 \, |y_n|^2_{\max} \, |G_n(x, \xi)|_{\max},$$

hence $|A_{1n}|$ is bounded by virtue of (2.264), (2.266).

3)

$$|A_2| \leqq h_2 l \, |y|^2_{\max}$$

$$+ \left| \int_0^l y(\xi) \int_0^l [h_1' y'(x) + h_1 y''(x) - g y'(x)] G(x, \xi)\, dx\, d\xi \right|$$

$$+ |(\gamma-1)| F |y'(l)| \cdot |G(l, \xi)|_{\max},$$

and

$$|A_2| \leqq h_2 l \, |y|^2_{\max}$$

$$+ l^2 \, |y|_{\max} \Big[|h_1'|_{\max} \, |y'|_{\max} \, |h_1|_{\max} + |y''|_{\max}$$

$$+ g \, |y'|_{\max} \Big] \cdot |G|_{\max} + |\gamma-1| \cdot |F| \cdot |y'(l)| \cdot |G(l, \xi)|_{\max}.$$

According to assumptions (2.264), (2.265), (2.266), (2.270), it thus becomes apparent that $|A_2|$ is also bounded.

4)

$$|A_{1n}-A_1| \leqq \left|\int_0^l [y_n T_{1n} y_n - y T_{1n} y] d\xi\right| + \left|\int_0^l [y T_{1n} y - y T_1 y] d\xi\right|. \quad (2.284)$$

Let the first term on the right hand side be estimated. By means of (2.271), (2.275) it can be transformed into

$$\left|\int_0^l [y_n T_{1n} y_n - y T_{1n} y]\, d\xi\right|$$

$$= \left|\int_0^l [y_n(\xi) \int_0^l y_n(x) G_n(x,\xi)\, dx] d\xi - \int_0^l [y(\xi) \int_0^l y(x) G_n(x,\xi)\, dx] d\xi\right|$$

$$= \left|\int_0^l \int_0^l [y_n(\xi) y_n(x) - y(\xi) y(x)] G_n(x,\xi)\, dx\, d\xi\right|.$$

That leads to

$$\left|\int_0^l [y_n T_{1n} y_n - y T_{1n} y] d\xi\right| \leqq \int_0^l \int_0^l |y_n(\xi) y_n(x) - y(\xi) y(x)| \cdot |G_n(x,\xi)|\, dx\, d\xi$$

$$\leqq |G_n(x,\xi)|_{\max} \int_0^l \int_0^l \{|y_n(\xi)[y_n(x)-y(x)]| + |y(x)[y_n(\xi)-y(\xi)]|\} dx\, d\xi$$

$$\leqq |G_n|_{\max} \{\int_0^l |y_n(\xi)|\, d\xi \cdot \int_0^l |y_n(x)-y(x)|\, dx + \int_0^l |y(x)|\, dx \cdot \int_0^l |y_n(\xi)-y(\xi)|\, d\xi\}$$

$$\leqq |G_n|_{\max}\, l^2 (|y_n|_{\max} + |y|_{\max}) \cdot |y_n - y|_{\max}.$$

Since G_n, y_n and y are bounded, and since y_n converges toward y according to (2.247), the last inequality yields

$$\left|\int_0^l [y_n T_{1n} y_n - y T_{1n} y] d\xi\right| \leqq \frac{1}{2}\, \epsilon \text{ for } n > N. \quad (2.285)$$

Now, let the second term on the right hand side be estimated which reads

$$\left| \int_0^l [yT_{1n}y - yT_1 y] d\xi \right|$$

$$= \left| \int_0^l y(\xi) \int_0^l y(x) G_n(x, \xi) dx\, d\xi - \int_0^l y(\xi) \int_0^l y(x) G(x, \xi)\, dx\, d\xi \right|$$

$$= \left| \int_0^l \int_0^l [y(\xi) y(x) G_n(x, \xi) - y(\xi) y(x) G(x, \xi)] dx\, d\xi \right| .$$

This relationship yields

$$\left| \int_0^l [yT_{1n}y - yT_1 y] d\xi \right| \leqq \int_0^l \int_0^l |y(\xi) y(x)| \cdot |G_n(x, \xi) - G(x, \xi)|\, dx\, d\xi$$

$$\leqq |G_n - G|_{\max}\, l^2\, |y|^2_{\max} .$$

Since y is bounded and G_n tends toward G, the result is

$$\left| \int_0^l [yT_{1n}y - yT_1 y] d\xi \right| \leqq \frac{1}{2}\, \epsilon \text{ for } n > N. \tag{2.286}$$

Interserting (2.285) and (2.286) in (2.284) yields

$$|A_{2n} - A_1| \leqq \epsilon \text{ for } n > N. \tag{2.287}$$

5)

$|A_2 - A_{2n}|$

$$= \left| h_2 \int_0^l y^2(\xi) d\xi - \int_0^l y(\xi) T_2 y(\xi) d\xi - h_2 \int_0^l y_n^2(\xi) d\xi + \int_0^l y_n(\xi) T_{2n} y_n(\xi) d\xi \right|$$

$$\leqq h_2 \left| \int_0^l y^2(\xi) d\xi - \int_0^l y_n^2(\xi) d\xi + \int_0^l y_n(\xi) T_{2n} y_n(\xi) d\xi - \int_0^l y(\xi) T_{2n} y(\xi) d\xi \right|$$

$$+ \left| \int_0^l y(\xi) T_{2n} y(\xi) d\xi - \int_0^l y(\xi) T_2 y(\xi) d\xi \right| . \tag{2.288}$$

The three terms on the right hand side will now be evaluated separately. For the first term it can be stated that

$$\left|\int_0^l y^2(\xi)\,d\xi - \int_0^l y_n^2(\xi)\,d\xi\right| = \left|\int_0^l y^2(\xi) - y_n^2(\xi)\;d\xi\right|$$

$$\leqq \int_0^l |\,y^2(\xi) - y_n^2(\xi)\,|\,d\xi \leqq \int_0^l |(y+y_n)(y-y_n)\,|\,d\xi$$

$$\leqq \int_0^l |y+y_n\,|\cdot|y-y_n\,|\,d\xi \leqq l(\,|y\,|_{\max} + |y_n\,|_{\max})\,|y-y_n\,|_{\max}.$$

According to (2.266) the quantities y and y_n are bounded. Moreover, y_n converges toward y. Hence,

$$h_2\left|\int_0^l y^2(\xi)\,d\xi - \int_0^l y_n^2(\xi)\,d\xi\right| \leqq \frac{1}{3}\;\epsilon \text{ for } n > N. \tag{2.289}$$

The second term on the right hand side of (2.288) reads

$$\left|\int_0^l y_n(\xi)\,T_{2n}y_n(\xi)\,d\xi - \int_0^l y(\xi)\,T_{2n}y(\xi)\,d\xi\right|$$

$$= \left|\int_0^l y_n(\xi)\int_0^l [(h_1 y_n'(x))' - gy_n'(x)]G_n(x,\xi)\,dx\,d\xi\right.$$

$$-(\gamma-1)Fy_n'(l)\int_0^l G_n(l,\xi)y_n(\xi)\,d\xi$$

$$-\int_0^l y(\xi)\int_0^l [(h_1 y'(x))' - gy'(x)]G_n(x,\xi)\,dx\,d\xi$$

$$\left.+(\gamma-1)Fy'(l)\int_0^l G_n(l,\xi)y(\xi)\,d\xi\right|.$$

Using

$$\phi^*(x) = (h_1 y'(x))' - gy'(x), \quad \phi_n^*(x) = (h_1 y_n'(x))' - gy_n'(x), \tag{2.290}$$

the previous equation leads to the inequality

$$\left|\int_0^l y_n(\xi)T_{2n}y_n(\xi)\,d\xi-\int_0^l y(\xi)T_{2n}y(\xi)\,d\xi\right| \tag{2.291}$$

$$\leqq\left|\int_0^l y_n(\xi)\int_0^l \phi_n^*(x)G_n(x,\xi)\,dx\,d\xi-\int_0^l y(\xi)\int_0^l \phi^*(x)G_n(x,\xi)\,dx\,d\xi\right|$$

$$+\left|(\gamma-1)Fy'(l)\int_0^l G_n(l,\xi)y(\xi)\,d\xi-(\gamma-1)Fy_n'(l)\int_0^l G_n(l,\xi)y_n(\xi)\,d\xi\right|.$$

The first term on the right hand side of (2.291) yields

$$\left|\int_0^l\int_0^l [y_n(\xi)\phi_n^*(x)G_n(x,\xi)-y(\xi)\phi^*(x)G_n(x,\xi)]\,dx\,d\xi\right|$$

$$=\left|\int_0^l\int_0^l [y_n(\xi)\phi_n^*(x)-y(\xi)\phi^*(x)]G_n(x,\xi)\,dx\,d\xi\right|$$

$$\leqq|G_n|_{\max}\int_0^l\int_0^l |y_n(\xi)\phi_n^*(x)-y(\xi)\phi^*(x)|\,dx\,d\xi \tag{2.292}$$

$$\leqq|G_n|_{\max}\int_0^l\int_0^l [\,|y_n(\xi)(\phi_n^*(x)-\phi^*(x)|+|\phi^*(x)(y_n(\xi)-y(\xi))|\,]\,dx\,d\xi$$

$$\leqq|G_n|_{\max}\{l\,|y_n|_{\max}\int_0^l [\,|\phi_n^*(x)-\phi^*(x)|\,]\,dx+l^2\,|\phi^*|_{\max}\,|y_n-y|_{\max}\}.$$

Moreover,

$$\int_0^l|\phi_n^*(x)-\phi^*(x)|\,dx\leqq\int_0^l|(h_1y_n')'-(h_1y')'|\,dx+g\int_0^l y_n'-y'|\,dx$$

$$\leqq\int_0^l|(h_1y_n')'-(h_1y')'|\,dx+gl\,|y_n'-y'|_{\max}.$$

In addition,

$$\int_0^l |(h_1 y_n')' - (h_1 y')'| dx$$

$$= \left| \int_0^{a_1} [(h_1 y_n')' - (h_1 y')'] dx \right| + \ldots + \left| \int_{a_{M-1}}^{a_M} [(h_1 y_n')' - (h_1 y')'] dx \right| ,$$

where $x = a_1, x = a_2, \ldots, x = a_M$ are the absissae of the zeros of $(h_1 y_n')' - (h_1 y')'$. As a consequence,

$$\int_0^l |(h_1 y_n')' - (h_1 y')'| dx = \left| h_1 y_n' - h_1 y' \right|_0^{a_1} + \ldots + \left| h_1 y_n' - h_1 y' \right|_{a_{M-1}}^{a_M}$$

$$\leqq (M+2) |h_1|_{max} |y_n' - y'|_{max} .$$

Using this inequality above yields

$$\int_0^l \phi_n^*(x) - \phi^*(x) | dx \leqq (M+2) |h_1|_{max} |y_n' - y'|_{max} + gl |y_n' - y'|_{max} . \tag{2.293}$$

Furthermore,

$$|\phi^*|_{max} \leqq |h_1|_{max} |y''|_{max} + |h_1'|_{max} |y'|_{max} + g |y'|_{max} . \tag{2.294}$$

By virtue of (2.293) and (2.294), inequality (2.292) changes into

$$\left| \int_0^l \int_0^l [y_n(\xi) \phi_n^*(x) G_n(x, \xi) - y(\xi) \phi^*(x) G_n(x, \xi)] dx \, d\xi \right|$$

$$\leqq |G_n|_{max} \{ l |y_n|_{max} [(M+2) |h_1|_{max} + gl] |y_n' - y'|_{max}$$

$$+ l^2 [|h_1|_{max} |y''|_{max} + |h_1'|_{max} + g |y'|_{max}] |y_n - y|_{max} \} .$$

Since y_n, y', y'', h_1, h_1' and G_n are bounded according to (2.264), (2.265), (2.266), (2.269), (2.270), M is a finite integer, and y_n and y_n' converge toward y and y', respectively,

$$\left|\int_0^l\int_0^l [y_n(\xi)\phi_n^*(x)G_n(x,\xi)-y(\xi)\phi^*(x)G_n(x,\xi)]\,dx\,d\xi\right| \leqq \frac{1}{6}\,\epsilon \text{ for } n>N. \tag{2.295}$$

The second term on the right hand side of (2.291) yields

$$\left|(\gamma-1)F_y'(l)\int_0^l G_n(l,\xi)y(\xi)\,d\xi-(\gamma-1)Fy_n'(l)\int_0^l G_n(l,\xi)y_n(\xi)\,d\xi\right|$$

$$\leqq|\gamma-1|F\left|[y'(l)-y_n'(l)]\int_0^l G_n(l,\xi)y(\xi)\,d\xi\right.$$

$$\left.+y_n'(l)\left[\int_0^l G_n(l,\xi)y(\xi)\,d\xi-\int_0^l G_n(l,\xi)y_n(\xi)\,d\xi\right]\right|$$

$$\leqq|\gamma-1|F\{|y'(l)-y_n'(l)|\,l\,|y|_{\max}|G_n|_{\max}+|y_n'(l)|\,l\,|y-y_n|_{\max}|G_n|_{\max}\}$$

$$\leqq|\gamma-1|F|G_n|_{\max}l\{|y|_{\max}|y'-y_n'|_{\max}+|y_n'|_{\max}|y-y_n|_{\max}\}.$$

According to (2.264), (2.266), (2.270), quantities G_n, y, y_n' are bounded. In addition, y_n' converges toward y' and y_n toward y. Therefore,

$$\left|(\gamma-1)Fy'(l)\int_0^l G_n(l,\xi)y(\xi)\,d\xi-(\gamma-1)Fy_n'(l)\int_0^l G_n(l,\xi)y_n(\xi)\,d\xi\right|$$

$$\leqq \frac{1}{6}\,\epsilon \text{ for } n>N. \tag{2.297}$$

Inserting (2.295), (2.296) in (2.291) yields

$$\left|\int_0^l y_n(\xi)T_{2n}y_n(\xi)\,d\xi-\int_0^l y(\xi)T_{2n}y(\xi)\,d\xi\right| \leqq \frac{1}{3}\,\epsilon \text{ for } n>N. \tag{2.297}$$

The third term on the right hand side of (2.288) yields by means of (2.272), (2.276), and (2.290)

$$\left|\int_0^l y(\xi)T_{2n}y(\xi)\,d\xi-\int_0^l y(\xi)T_2y(\xi)\,d\xi\right|$$

$$= \left| \int_0^l y(\xi) \int_0^l \phi^*(x) G_n(x,\xi)\,dx\,d\xi - \int_0^l y(\xi) \int_0^l \phi^*(x) G(x,\xi)\,dx\,d\xi \right.$$

$$\left. +(\gamma-1)Fy'(l) \int_0^l G(l,\xi) y(\xi)\,d\xi - (\gamma-1)Fy'(l) \int_0^l G_n(l,\xi) y(\xi)\,d\xi \right|$$

$$= \left| \int_0^l y(\xi) \int_0^l \phi^*(x) [G_n(x,\xi) - G(x,\xi)]\,dx\,d\xi \right.$$

$$\left. +(\gamma-1)Fy'(l) \int_0^l y(\xi)[G(x,\xi) - G_n(x,\xi)]\,d\xi \right|$$

$$\leqq \int_0^l \int_0^l y(\xi)\phi^*(x) | \cdot | G_n(x,\xi) - G(x,\xi) |\,dx\,d\xi$$

$$+ |\gamma-1| F |y'(l)| \int_0^l |y(\xi)| \cdot |G_n(x,\xi) - G(x,\xi)|\,d\xi$$

$$\leqq |G_n - G|_{\max} \left\{ l\,|y|_{\max} \left(l\,|\phi^*|_{\max} + |\gamma-1|\,F\,|y'|_{\max} \right) \right\}$$

$$\leqq |G_n - G|_{\max}\, l\,|y|_{\max} \left[l \left(|h_1'|_{\max} |y'|_{\max} \right. \right.$$

$$\left. \left. + |h_1|_{\max} |y''|_{\max} + g\,|y'|_{\max} \right) + |\gamma-1|\,F\,|y'|_{\max} \right].$$

According to (2.265), (2.266), (2.269), (2.270), all absolute values involved in the preceding inequality are bounded. Moreover, G_n converges toward G. Therefore,

$$\left| \int_0^l y(\xi) T_{2n} y(\xi)\,d\xi - \int_0^l y(\xi) T_2 y(\xi)\,d\xi \right| \leqq \frac{1}{3}\,\epsilon \text{ for } n > N \tag{2.298}$$

results from that inequality.

Using (2.289), (2.297), (2.298) in (2.288) yields

$$|A_2 - A_{2n}| \leqq \epsilon \text{ for } n > N. \tag{2.299}$$

Comparing the results in 1), 2), 3), and the inequalities derived in 4) and 5) with the conditions of convergence (2.281), (2.282), (2.283), one realizes that these conditions are all satisfied. Thus, the convergence of λ_n toward λ, i.e., the convergence of the method of Ritz, has been proved for certain *nonconservative, nonselfadjoint problems* under the assumption that only *admissible* coordinate functions have been used.

The case where the coordinate functions are comparison functions which satisfy *all* boundary conditions can be treated correspondingly. The absolute and uniform convergence of the series (2.219), (2.247), (2.248) is without question for comparison functions. Everything becomes even simpler as in the integral equations (2.256), (2.263) the second term on the right hand side always vanishes. In the same manner, the operators T_2 and T_{2n} defined by (2.272), (2.276) become simpler. Since everything else proceeds in the same way as before, it has not to be repeated explicitly. It can immediately be concluded that *if the method of Ritz converges using admissible coordinate functions it certainly converges also when comparison functions are being used as coordinate functions.*

2.3 Examples

Firstly, the rod problems shown in Figure 18 may be considered. For these problems,

$$
\begin{aligned}
&Tw = \mu \ddot{w} \\
&h_0 = 0, \\
&h_1 = -g(l-x), \\
&h_2 = \alpha, \\
&Q_x[w] = \epsilon g w', \qquad \epsilon = 0 \text{ in cases } K1 \text{ to } K4, \\
&\qquad\qquad\qquad\quad\;\; \epsilon = 1 \text{ in cases } NK1 \text{ to } NK4.
\end{aligned}
\tag{2.300}
$$

Hence, (2.145) reads

$$
\begin{aligned}
&Tw + \sum_{k=0}^{2} (-1)^k [h_k w^{(k)}]^{(k)} + \epsilon g w' = 0, \\
&(Gw)_R = 0
\end{aligned}
$$

or

$$
\begin{aligned}
&\mu \ddot{w} + g[(l-x)w']' + \alpha w^{\mathrm{IV}} + \epsilon g w' = 0, \\
&(Gw)_R = 0,
\end{aligned}
\tag{2.300}
$$

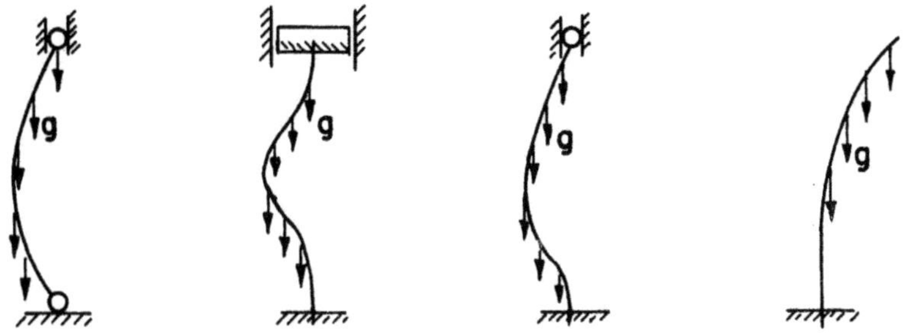

Figure 18a. Buckling of elastic rods subjected to a conservative load, $\epsilon = 0$, *K*1 *up to K*4

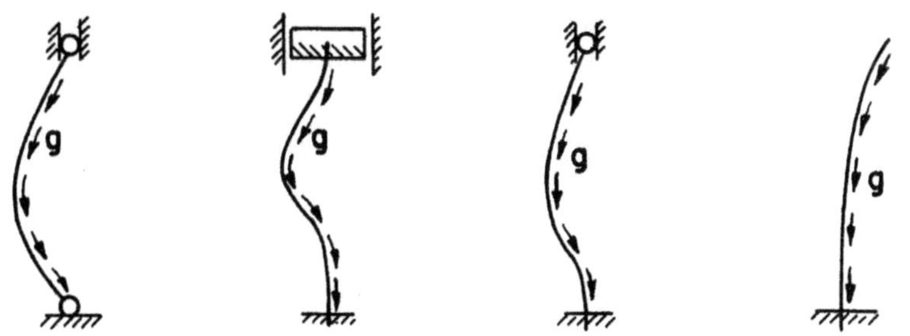

Figure 18b. Buckling of elastic rods subjected to a nonconservative load, $\epsilon = 1$, *NK*1 *up to Nk*4

respectively. By means of (1.100), and (1.101), (2.301) changes into

$$-\lambda y + \alpha y^{IV} + g[(l-x)y']' + \epsilon g y' = 0,$$

$$(Gy)_R = 0, \tag{2.302}$$

which is to be compared with (2.146). It follows that

$$S_1 y = \alpha y^{IV},$$

$$S_2 y = [(l-x)y']',$$

$$Ny = y'. \tag{2.303}$$

As boundary conditions $(Gy)_R = 0$, equations (2.48) for $K1, NK1$, equations (2.49) for $K2, NK2$, equations (2.50) for $K3, NK3$, and equations (2.51) for $K4$, $NK4$ have to be used. The eigenfunctions of the auxiliary problem (1.113), (1.114), which are obtained when $H \equiv S_1$, shall be used as coordinate functions. With $\lambda_i/\alpha \equiv \omega_i$, this auxiliary problem reads

$$\phi_i^{\mathrm{IV}} = \omega_i \phi_i,$$
$$(G\phi_i)_R = 0, \tag{2.304}$$

where

$$\begin{aligned} (G\phi_i)_R &= \phi_i(0) = \phi_i''(0) = \phi_i\,(l) = \phi_i''(l) = 0 \quad \text{for } K1, NK1, \\ &= \phi_i(0) = \phi_i'\,(0) = \phi_i\,(l) = \phi_i'\,(l) = 0 \quad \text{for } K2, NK2, \\ &= \phi_i(0) = \phi_i'\,(0) = \phi_i\,(l) = \phi_i''(l) = 0 \quad \text{for } K3, NK3, \\ &= \phi_i(0) = \phi_i'\,(0) = \phi_i''(l) = \phi_i'''(l) = 0 \quad \text{for } K4, NK4. \end{aligned} \tag{2.305}$$

Problem (2.216), (2.217) is an example. The eigenfunctions of the free vibrations of the unloaded rods are solutions to (2.304), (2.305). They are tabulated in [29]. Let l be the length of the rod. Then, after normalization,

$$\phi_i = \sqrt{\frac{2}{l}}\,\sin\frac{i\pi x}{l} = \sqrt{\frac{2}{l}}\,\sin \lambda_i x,\ \lambda_i = \frac{i\pi}{l} \ \text{for } K1, NK1. \tag{2.306}$$

In all other cases,

$$\phi_i = \sqrt{\frac{1}{l}}\left[\cos h\ \lambda_i x - \cos\lambda_i x - \sigma_i\,(\sin h\ \lambda_i x - \sin\lambda_i x)\right], \tag{2.307}$$

where

Table 2.1

	$K2,\ NK2$	$K3,\ NK3$	$K4,\ NK4$
$\lambda_1 l$	4.730	3.927	1.875
$\lambda_2 l$	7.853	7.069	4.694
σ_1	0.983	1.001	0.734
σ_2	1.001	1.000	1.018

(2.308)

If (2.303), (2.304), (2.306), and (2.307) are used, it can be shown that

$$\int_0^l \phi_i S_1 \phi_j dx = \alpha \int_0^l \phi_i \phi_j^{\mathrm{IV}}\, dx = \alpha\omega_i\delta_{ij}, \quad \omega_i = \lambda_i^4. \tag{2.309}$$

Further data which are needed for subsequent calculation are contained in Table 2.2.

Table 2.2

$\frac{1}{\alpha}\int_0^l \phi_1 S_1 \phi_1 dx$	$\pi^4 l^{-4}$	$500.56l^{-4}$	$237.72l^{-4}$	$12.36l^{-4}$
$\frac{1}{\alpha}\int_0^l \phi_2 S_1 \phi_2 dx$	$16\pi^4 l^{-4}$	$3803.54l^{-4}$	$2496.49l^{-4}$	$485.52l^{-4}$
$\int_0^l \phi_1 S_2 \phi_1 dx$	$-4.94l^{-1}$	$-6.15l^{-1}$	$-4.29l^{-1}$	$-1.57l^{-1}$
$\int_0^l \phi_2 S_2 \phi_1 dx$	$-4.45l^{-1}$	$-3.340622l^{-1}$	$-2.08l^{-1}$	$0.42l^{-1}$
$\int_0^l \phi_1 S_2 \phi_2 dx$	$-4.45l^{-1}$	$-3.340622l^{-1}$	$-2.08l^{-1}$	$0.42l^{-1}$
$\int_0^l \phi_2 S_2 \phi_2 dx$	$-19.74l^{-1}$	$-23.03l^{-1}$	$-18.41l^{-1}$	$-8.65l^{-1}$
$\int_0^l \phi_1 N \phi_1 dx$	0	0	0	$2.00l^{-1}$
$\int_0^l \phi_2 N \phi_1 dx$	$2.67l^{-1}$	$3.34l^{-1}$	$3.00l^{-1}$	$0.76l^{-1}$
$\int_0^l \phi_1 N \phi_2 dx$	$-2.67l^{-1}$	$-3.34l^{-1}$	$-3.00l^{-1}$	$-4.76l^{-1}$
$\int_0^l \phi_2 N \phi_2 dx$	0	0	0	$2.00l^{-1}$

The coordinate functions ϕ_i in (2.306) and (2.307) satisfy *all* boundary conditions. Therefore, the variational principle may be used in its special form (2.2143). It has already been shown that the first line in (2.144) corresponds to the first line in (2.145). However, it has to be noted that $Q_x[w] = \epsilon g w'$, where $\epsilon = 1$ or $\epsilon = 0$. Finally, comparing with the integrand of (2.143) yields

$$\int_T \int_0^l \{Tw + \sum_{k=1}^{2} (-1)^k [h_k w^{(k)}]^{(k)} + \epsilon g w'\} \delta w \, dx \, dt = 0. \tag{2.310}$$

By means of (1.100) and (1.101), equation (2.310) can be transformed into

$$\int_0^l \{-\lambda y + \sum_{k=0}^{2} (-1)^k [h_k y^{(k)}]^{(k)} + \epsilon g y'\} \delta y \, dx \, dt = 0. \tag{2.311}$$

Applying (2.300) with respect to functions h_i yields

$$\int_0^l \{-\lambda y + \alpha y^{\mathrm{IV}} + g[(l-x)y']' + \epsilon g y'\} \delta y \, dx = 0. \tag{2.312}$$

Because of (2.303), the previous equation is equivalent to

$$\int_0^l \{-\lambda y + S_1 y + g S_2 y + g\epsilon N y\} \delta y \, dx = 0. \tag{2.313}$$

Applying (1.111) to (2.313) and using two coordinate functions only yields

$$\int_0^l \{-\lambda(a_1\phi_1 + a_2\phi_2) + S_1(a_1\phi_1 + a_2\phi_2) + g S_2(a_1\phi_1 + a_2\phi_2)$$

$$+ g\epsilon N(a_1\phi_1 + a_2\phi_2)\}(\delta a_1 \phi_1 + \delta a_2 \phi_2) \, dx = 0.$$

Since this is supposed to hold for arbitrary variations δa_1, δa_2, the following equations

$$\begin{aligned}
&a_1 \left[-\lambda + \int_0^l \phi_1 S_1 \phi_1 \, dx + g\left(\int_0^l \phi_1 S_2 \phi_1 \, dx + \epsilon \int_0^l \phi_1 N \phi_1 \, dx\right)\right] \\
&\quad + a_2 \left[\phi_1 S_1 \phi_2 dx + g\left(\int_0^l \phi_1 S_2 \phi_2 dx + \epsilon \int_0^l \phi_1 N \phi_2 dx\right)\right] = 0, \\
&a_1 \left[\int_0^l \phi_2 S_1 \phi_1 dx + g\left(\int_0^l \phi_2 S_2 \phi_1 dx + \epsilon \int_0^l \phi_2 N \phi_1 dx\right)\right] \\
&\quad + a_2 \left[-\lambda + \int_0^l \phi_2 S_1 \phi_2 dx + g\left(\int_0^l \phi_2 S_2 \phi_2 dx + \epsilon \int_0^l \phi_2 N \phi_2 dx\right)\right] = 0
\end{aligned} \tag{2.314}$$

must be satisfied. These are the so called Galerkin equations. Due to (2.309), and using the abbreviation

$$\phi_{ij} = \int_0^l \phi_i S_2 \phi_j dx + \epsilon \int_0^l \phi_i N \phi_j dx, \tag{2.315}$$

they may be rewritten in a more concise form as

$$a_1 \left[-\lambda + \alpha\lambda_1^4 + g\phi_{11}\right] + a_2 g\phi_{12} = 0,$$

$$a_1 g\phi_{21} + a_2 \left[-\lambda + \alpha\lambda_2^4 + g\phi_{22}\right] = 0. \tag{2.316}$$

Since these equations are homogeneous,

$$\begin{vmatrix} -\lambda + \alpha\lambda_1^4 + g\phi_{11} & g\phi_{12} \\ g\phi_{21} & -\lambda + \alpha\lambda_2^4 + g\,\phi_{22} \end{vmatrix} = 0 \tag{2.317}$$

must hold. A consequence of this condition is the eigenvalue equation

$$\lambda^2 - \lambda\left[\alpha(\lambda_1^4 + \lambda_2^4) - \lambda g(\phi_{11} + \phi_{22}) + g\left[\alpha(\lambda_1^4\phi_{22} + \lambda_2^4\phi_{11})\right]\right.$$

$$+ g^2(\phi_{11}\phi_{22} - \phi_{12}\phi_{21}) + \alpha^2\lambda_1^4\lambda_2^4 = 0. \tag{2.318}$$

A solution of (2.318) can be obtained from case to case using (2.315) and the values in Table 2.2. The graphical representation of (2.318) can be found in Figures 19 and 20. The trace of the eigenvalue curves indicates that systems $K1$ to $K4$ as well as systems $NK1$ and $NK2$ are divergence type systems. System $NK3$ is a pseudo-divergence type system, and $NK4$ is a flutter type system. The smallest critical loads shown in the figures are also tabulated in Table 2.3.

Table 2.3

Critical Loads $g_{crit}\, l^3/\alpha_{,min}$				
Support				
Conservative Load	18.57	74.63	53.50	7.85
Type of Instability	Divergence	Divergence	Divergence	Divergence
Nonconservative Load	18.956	80.255	57.01	40.05
Type of Instability	Divergence	Divergence	Divergence	Flutter

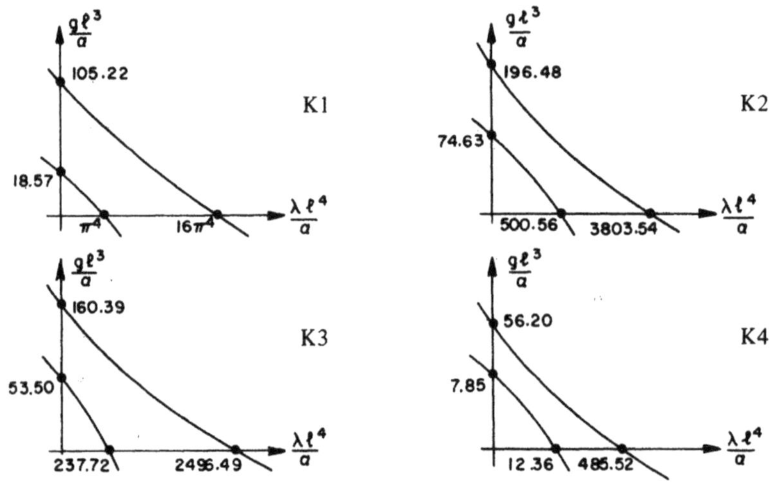

Figure 19. Eigenvalue curves of cases K1 to K4 in Figure 18a, $\epsilon = 0$

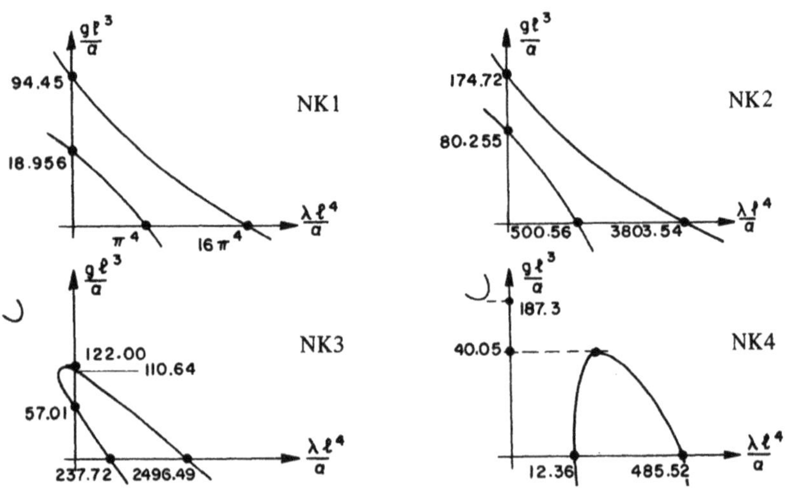

Figure 20. Eigenvalue curves of cases NK1 to NK4 in Figure 18b, $\epsilon = 1$

As can be seen, in all cases the critical loads at conservative loading are lower bounds to the critical loads at nonconservative loading. It can easily be shown in the case of completely supported divergence and pseudo-divergence type systems that this must in fact be the case:

For systems $K1$ to $K3$ and $NK1$ to $NK3$, corresponding conservative and nonconservative systems have equal boundary conditions. Hence, the eigenfunctions y_{nk} of the three nonconservative systems $NK1$ to $NK3$ can serve as comparison functions for the three conservative systems $K1$ to $K3$. Thus, Rayleigh's quotient (2.39) can be written in terms of y_{nk} to yield

$$\bar{\lambda} = \frac{\int_0^l [y_{nk} S_1 y_{nk} + g y_{nk} S_2 y_{nk}] dx}{\int_0^l y_{nk}^2 \, dx}. \tag{2.319}$$

However,

$$\lambda_k \leqq \bar{\lambda} \tag{2.320}$$

Holds true for the actual eigenvalue λ of the conservative systems, $K1$ to $K3$, as has been shown in Section 2.1. For nonconservative systems $NK1$ to $NK3$, quotient (2.150) may be formed which yields the actual eigenvalue λ_{NK} of the nonconservative systems if $y \equiv y_{nk}$. Hence,

$$\lambda_{NK} = \frac{\int_0^l [y_{nk} S_1 y_{nk} + g y_{nk} S_2 y_{nk} + g y_{nk} N y_{nk}] dx}{\int_0^l y_{nk}^2 \, dx}. \tag{2.321}$$

The systems $NK1$ to $NK3$ are completely supported i.e.,

$$y_{nk}(0) = y_{nk}(l) = 0. \tag{2.322}$$

According to the third line in (2.303),

$$\int_0^l y_{nk} N y_{nk} \, dx = \int_0^l y_{nk} y'_{nk} \, dx. \tag{2.323}$$

Integration by parts yields

$$\int_0^l y_{nk} N y_{nk}\, dx = \frac{1}{2}\left[y_{nk}^2\right]_0^l = 0 \tag{2.324}$$

by virtue of (2.322). Using (2.324) in (2.321) and comparing with (2.319) yields

$$\lambda_{NK} = \frac{\int_0^l [y_{nk} S_1 y_{nk} + g y_{nk} S_2 y_{nk}]dx}{\int_0^l y_{nk}^2\, dx} \equiv \bar{\lambda}. \tag{2.325}$$

Combining (2.320) with (2.325) yields

$$\lambda_K \leqq \lambda_{NK}. \tag{2.326}$$

Thus, the eigenvalues λ_{NK} of completely supported systems with a uniformly distributed nonconservative load are an upper bound to the eigenvalues λ_K of the same system subjected to a conservative load.

Assuming in addition that the systems under consideration behave like divergence type systems at least up to the smallest critical load $g_{\mathrm{crit,min}}$, the trace of the eigenvalue curve of the nonconservative case is always located to the right of the eigenvalue curve of the corresponding conservative case. Since one is dealing with divergence or at least with pseudo-divergence, both curves intersect the g-axis. Necessarily, the curve to the right intersects at a point which is above of the point at which the curve to the left intersects the g-axis. The ordinates of the points of intersection represent the critical loads. Let $g_{\mathrm{crit,min},K}$ be the smallest critical divergence load of the nonconservative case, then, obviously,

$$g_{\mathrm{crit,min},K} < g_{\mathrm{crit,min},NK} \tag{2.327}$$

must hold. Consequently, the critical load of the conservative case is a lower bound to the critical load of the corresponding nonconservative case.

Table 2.3 yields additional information. In Section 2.2 it was predicted using a perturbation technique that for the completely supported systems $K1$ to $K3$ and $NK1$ to $NK3$ the corresponding eigenvalues will differ only with respect to the second order of the perturbation parameter. That means that the eigenvalue curves of corresponding systems, for example $K1$ and $NK1$ etc., should be close to each other in the g, λ-plane. But then, the points of intersection of the eigenvalue curves with the g-axis must be close together. Hence, only a small difference between $g_{\mathrm{crit,min},K}$ and $g_{\mathrm{crit,min},NK}$ has to be expected. This is indeed satisfied for $K1$ to $K3$ and $NK1$ to $NK3$ as a glance to Table 2.3 proves.

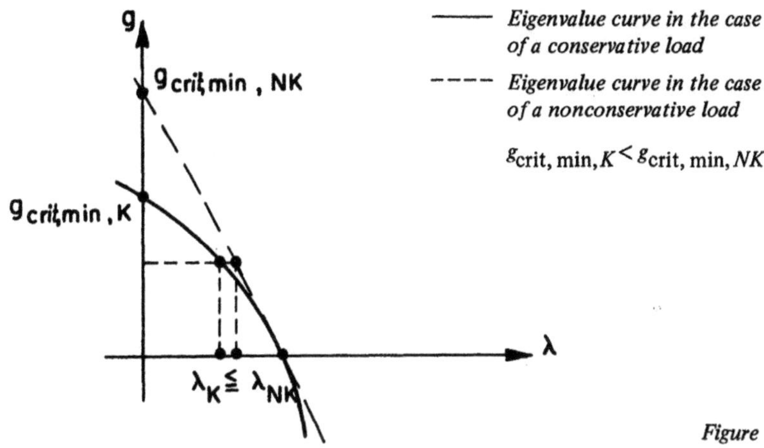

Figure 21

Now, it will be shown that the systems in Figure 18 can be classified without the need to solve (2.318) completely or to actually draw the eigenvalue curve. For this purpose, (2.191) may be used.

In the cases $K1$ to $K4$, the operator N is not present, since $\epsilon \equiv 0$, i.e., no nonconservative loads are present. Therefore,

$$\mathscr{P} = \left[\int_0^l \phi_1 S_2 \phi_2 \, dx\right] \cdot \left[\int_0^l \phi_2 S_2 \phi_1 \, dx\right].$$

However, S_2 is by definition a selfadjoint operator so that

$$\mathscr{P} = \left[\int_0^l \phi_1 S_2 \phi_2 \, dx\right]^2 > 0.$$

But then, the branches of the corresponding eigenvalue curves are located in domains I and II of Figure 11. The flutter line is not intersected, as one has to deal with the case displayed in Figure 12. Hence, the systems concerned are divergence type systems. The intersection of the straight line $[\phi_1] = 0$, i.e., of

$$g = -\frac{C_{1\phi_1}}{C_{2\phi_1}} + \frac{\lambda}{C_{1\phi_1}}, \tag{2.185$_1$}$$

with the g-axis yields

$$\tilde{g}_{1\text{crit}} = -\frac{C_{1\phi_1}}{C_{2\phi_1}} \tag{2.328}$$

as an upper bound for the smallest critical load $g_{1\text{crit}} = g_{\text{crit,min}}$.

By means of (2.186) and by virtue of (2.324), which holds true not only for y_{nk} but for ϕ_1 as well, one arrives at

$$\tilde{g}_{1\text{crit}} = -\frac{\int_0^l \phi_1 S_1 \phi_1 \, dx}{\int_0^l \phi_1 S_2 \phi_1 \, dx} \,. \tag{2.329}$$

This expression is calculated using Table 2.2. The results are given in Table 2.4.

Table 2.4

	$K1$	$K2$	$K3$	$K4$
$\frac{l^3}{\alpha}\tilde{g}_{1\text{crit}}$	19.7	81.39	55.41	7.87
$\frac{l^3}{\alpha} g_{\text{crit,min}}$	18.57	74.63	53.50	7.85

One has obtained in this way practically useful approximations $\tilde{g}_{1\text{crit}}$ of the actual critical loads $g_{\text{crit,min}}$, and one is relieved of the pain of determining the true eigenvalue curves.

In the cases $NK1$ to $NK4$,

$$\mathscr{P} = \left[\int_0^l \phi_1 S_2 \phi_2 \, dx + \int_0^l \phi_1 N \phi_2 \, dx\right] \cdot \left[\int_0^l \phi_2 S_2 \phi_1 \, dx + \int_0^l \phi_2 N \phi_1 \, dx\right], \tag{2.330}$$

according to (2.191). For $NK1$ follows from (2.330) and Table 2.2:

$$\mathscr{P} = [-4.45 - 2.67] \cdot [-4.45 + 2.67] l^{-2},$$

$$= 7.12 \cdot 1.78 l^{-2} = 8.66 l^{-2} > 0.$$

Hence, one has to deal with a divergence type system corresponding to that in Figure 12. Again, $\mathscr{E}\ [\phi] = 0$ yields an approximation for the buckling load. Since $NK1$ is completely supported, (2.324) is valid for ϕ_1, and from (2.328), relationship (2.329) is obtained by means of (2.186). Using the values in Table 2.2 yields

$$\frac{l^3}{\alpha}\,\tilde{g}_{1\mathrm{crit}} = \frac{\int_0^l \phi_1 S_1 \phi_1\,dx}{\int_0^l \phi_1 S_2 \phi_1\,dx}\,\frac{l^3}{\alpha} = \frac{\pi^4}{4.94} = 19.7,$$

which indeed is an acceptable approximation of $(l^3/\alpha)\,g_{\mathrm{crit,min},NK} = 18.956$. For $NK2$ follows from (2.330) and Table 2.2:

$$\mathscr{P} = [-3.34 - 3.34]\cdot[-3.34062 + 3.34]\,l^{-2},$$

$$\mathscr{P} = 6.68 \cdot 0.62 \cdot 10^{-3} l^{-2} = 4.14 \cdot 10^{-3} l^{-2} > 0.$$

Therefore, $NK2$ is also a divergence type system. In addition, it is a completely supported system. Hence, by the same token as for $NK1$,

$$\frac{l^3}{\alpha}\,\tilde{g}_{1\mathrm{crit}} = \frac{\int_0^l \phi_1 S_1 \phi_1\,dx}{\int_0^l \phi_1 S_2 \phi_1\,dx}\,\frac{l^3}{\alpha}$$

is an approximation of $(l^3/\alpha)\,g_{\mathrm{crit,min},NK}$. Using the values in Table 2.2 yields

$$\frac{l^3}{\alpha}\,\tilde{g}_{1\mathrm{crit}} = \frac{500.56}{6.15} = 81.39,$$

which is a good approximation of

$$\frac{l^3}{\alpha}\,g_{1\mathrm{crit,min},NK} = 80.255.$$

Now, let $NK3$ be investigated. For (2.330), the values in Table 2.2 yield:

$$\mathscr{P} = [-2.08 - 3.00]\cdot[-2.08 + 3.00]\,l^{-2} = -4.66\,l^{-2}.$$

Since $\mathscr{P} < 0$, the traces of the eigenvalue curve are located in domains III and IV of the g, λ-plane. Thus, the flutter line is intersected, and flutter is possible. Now, the traces of lines $\mathscr{E}\,[\phi_1] = 0$ and $\mathscr{E}\,[\phi_2] = 0$ may be checked. These lines are given by (2.185). Since according to (2.186) and Table 2.2

$$C_{2\phi_1} = \int_0^l \phi_1 S_2 \phi_1\,dx + \int_0^l \phi_1 N \phi_1\,dx = -4.29\,l^{-1} < 0,$$

line $\mathscr{E}\,[\phi_1] = 0$ has a negative slope. In the same manner one finds that

$$C_{2\phi_2} = \int_0^l \phi_2 S_2 \phi_2 \, dx + \int_0^l \phi_2 N \phi_2 \, dx = -2.08 l^{-1} < 0.$$

That means, line $\mathscr{E}\,[\phi_2] = 0$ has a negative slope too. Consequently, the trace of the eigenvalue curve corresponds to the situations shown in Figure 13 or 14. Checking condition (2.201) by means of (2.200), (2.186) and the values in Table 2.2,

$$\mathscr{S} = \frac{(-237.72 \cdot 18.41 + 4.29 \cdot 2796.49)^2}{4 \cdot 237.72 \cdot 2496.49} l^{-2} = 16.5 l^{-2}$$

and

$$\mathscr{P} = -4.66 l^{-2} > \mathscr{S} = -16.5 l^{-2}$$

is the result. Hence, $NK3$ is a pseudo-divergence type system as shown in Figure 13. An approximation of the smallest critical load follows from intersecting $\mathscr{E}\,[\phi_1] = 0$ with the g-axis. The values in Table 2.2 yield,

$$\frac{l^3}{\alpha} \tilde{g}_{1\text{crit}} = \frac{\int_0^l \phi_1 S_1 \phi_1 \, dx}{\int_0^l \phi_1 S_2 \phi_1 \, dx} \frac{l^3}{\alpha} = \frac{237.72}{4.29} = 55.41.$$

That is in fair agreement with $(l^3/\alpha)\, g_{\text{crit,min},NK} = 57.01$.

For $NK4$,

$$\mathscr{P} = [0.42 - 4.76] \cdot [0.42 + 0.76] l^{-2} = -5.12 l^{-2} < 0$$

according to (2.330) and the values in Table 2.2. Hence, one has to deal with a flutter type system, whose eigenvalue curve traces are located in domains III and IV, thus intersecting the flutter line. The slope of line $\mathscr{E}\,[\phi_1] = 0$ is determined by

$$C_{2\phi_1} = \int_0^l \phi_1 S_2 \phi_1 \, dx + \int_0^l \phi_1 N \phi_1 \, dx = -1.57 l^{-1} + 2.00 l^{-1} = 0.43 l^{-1} > 0$$

according to the values in Table 2.2. The slope is positive. The slope of line $\mathscr{E}\,[\phi_2] = 0$ is obtained correspondingly from

$$C_{2\phi_2} = \int_0^l \phi_2 S\phi_2 \, dx + \int_0^l \phi_2 N\phi_2 \, dx = -8.65l^{-1} + 2.00l^{-1} = -6.65l^{-1} < 0$$

and turns out to be negative. Hence, the situation is as indicated in Figure 15. Since *NK*4 is the last case to be considered in connection with rod problems, it becomes evident, as already mentioned, that at least for rods with uniformly distributed load and mass, the cases shown in Figures 16 and 17 are without practical importance.

Now, let the flutter load of *NK*4 be determined without using the eigenvalue curve at all. For that purpose, the flutter line (2.196), which reads here

$$\mathscr{E}[\phi_1] + \mathscr{E}[\phi_2] = 0, \tag{2.331}$$

is intersected with the pair of lines (2.205), which reads here

$$(\mathscr{E}[\phi_1])^2 + g^2 \mathscr{P} = 0, \tag{2.332}$$

according to the procedure outlined in Section 2.2. Using the values in Table 2.2, equation (2.331) changes by virtue of (2.173) into

$$-2\left(\frac{\lambda l^4}{\alpha}\right) - 6.22\left(\frac{gl^3}{\alpha}\right) + 497.88 = 0. \tag{2.333}$$

Correspondingly, (2.332) yields

$$\left[-\left(\frac{\lambda l^4}{\alpha}\right) + 0.43\left(\frac{gl^3}{\alpha}\right) + 12.36\right]^2 - 5.12\left(\frac{gl^3}{\alpha}\right)^2 = 0, \tag{2.334}$$

using Table 2.2 and $\mathscr{P} = -5.12l^{-2}$.

From (2.33) follows

$$\frac{\lambda l^4}{\alpha} = 248.94 - 3.11\left(\frac{gl^3}{\alpha}\right). \tag{2.335}$$

Using this expression in (2.334) yields

$$\left[-236.48 + 3.54\left(\frac{gl^3}{\alpha}\right)\right]^2 = 5.12\left(\frac{gl^3}{\alpha}\right)^2$$

or

$$-236.48 + 3.54\,\frac{gl^3}{\alpha} = \pm\ 2.26\,\frac{gl^3}{\alpha}.$$

Hence,

$$\frac{gl^3}{\alpha} = \frac{236.48}{5.8} = 40.7$$

and

$$\frac{gl^3}{\alpha} = \frac{236.48}{1.28} = 185.$$

These values agree favourably with the flutter loads 40.05 and 187.3 in Figure 20.

The next examples are concerned with plates. Firstly, the example in Figure 6 will be considered. It involves the free supported rectangular plate having the side ratio a/b and being subjected to a uniformly distributed, compressive follower force g, which is parallel to the edge having the length a. In Section 1.4 it has already been shown that in this case,

$$\frac{d}{dt}\left(\frac{\partial \mathcal{T}}{\partial w_t}\right) + \frac{\delta U}{\delta w} - Q_x[w] = \mu w_{tt} + D\Delta^2 w + g(a-x_1)w_{x_1x_1}. \tag{2.336}$$

The boundary conditions are given in (1.69), (1.70) and (1.71). Let

$$w(t, x_1, x_2) = f(t)y(x_1, x_2), \quad f(t) = e^{i\omega t}. \tag{2.337}$$

Then, the principle (2.143) yields

$$\int_0^a \int_0^b [-\lambda^* y + \Delta^2 y + g^*(a-x_1)y_{x_1x_1}]\delta y \, dx_1 \, dx_2 = 0 \tag{2.338}$$

according to (2.336) and (2.337). In (2.338),

$$\lambda^* = \frac{\mu\omega^2}{D}, \quad g^* = \frac{g}{D}. \tag{2.339}$$

An application of the principle in its form (2.143) requires coordinate functions satisfying all boundary conditions. Such functions are for example

$$\phi_{ij}(x_1, x_2) = \frac{2}{\sqrt{ab}} \sin\frac{i\pi x_1}{a} \sin\frac{j\pi x_2}{b}, \tag{2.340}$$

which makes

$$w_{ij}(t, x_1, x_2) = f(t)\phi_{ij}(x_1, x_2)$$

indeed satisfy (1.169), (1.70), and (1.71). In addition,

$$\int_0^a \int_0^b \phi_{ij}\phi_{kl}dx_1 fx_2 = \delta_{ik}\delta_{jl} \tag{2.341}$$

and

$$\Delta^2\phi_{ij} = \lambda_{ij}\phi_{ij}, \tag{2.342}$$

where

$$\lambda_{ij} = \left(\frac{i^2\pi^2}{a^2} + \frac{j^2\pi^2}{b^2}\right)^2. \tag{2.343}$$

These conditions can easily be verified.

Using

$$y(x_1, x_2) = \sum_i\sum_j a_{ij}\phi_{ij} \tag{2.344}$$

in (2.338) yields the system of equations

$$\sum_i\sum_j a_{ij}\left\{\int_0^a\int_0^b [-\lambda^*\phi_{ij}+\Delta^2\phi_{ij}+g^*(a-x_1)\phi_{ij,x_1x_1}]\phi_{kl}dx_1\,dx_2\right\} = 0,$$

$$k, l = 1,2,3,\ldots, \tag{2.345}$$

by virtue of the arbitrariness of the variations δa_{ij}. Using (2.341) in (2.345) and relationships

$$\int_0^a\int_0^b \Delta^2\phi_{ij}\phi_{kl}dx_1\,dx_2 = \lambda_{ij}\delta_{ik}\delta_{jl}, \tag{2.346}$$

$$\int_0^a\int_0^b \phi_{ij,x_1x_1}\phi_{kl}dx_1\,dx_2 = -\frac{i^2\pi^2}{a^2}\delta_{ik}\delta_{jl}, \tag{2.347}$$

which follows from (2.341) and (2.342), the following set of equations is obtained

$$\sum_i\sum_j a_{ij}\left\{\left[-\lambda^*+\lambda_{ij}-g^*\frac{i^2\pi^2}{a}\right]\delta_{ik}\delta_{jl}-g^*\int_0^a\int_0^b x_1\phi_{ij,x_1x_1}\phi_{kl}dx_1\,dx_2\right\} = 0,$$

$$k, l = 1,2,3,\ldots. \tag{2.348}$$

Let now the integral

$$\int_0^a \int_0^b x_1 \phi_{ij,\, x_1 x_1} \, \phi_{kl} dx_1 \, dx_2$$

$$= -\frac{i^2\pi^2}{a^2} \frac{4}{ab} \left[\int_0^a x_1 \sin\frac{i\pi x_1}{a} \sin\frac{k\pi x_1}{a} \, dx_1 \right] \cdot \left[\int_0^b \sin\frac{j\pi x_2}{b} \sin\frac{i\pi x_2}{b} dx_2 \right]$$

be evaluated. One arrives immediately at

$$\int_0^a \int_0^b x_1 \phi_{ij,\, x_1 x_1} \, \phi_{kl} dx_1 \, dx_2 = -\frac{i^2\pi^2}{a^2} \frac{2}{a} \delta_{ij} \int_0^a x_1 \sin\frac{i\pi x_1}{a} \sin\frac{k\pi x_1}{a} \, dx_1 . \tag{2.349}$$

Therefore, finally,

$$L_{i,k}^* = \int_0^a x_1 \sin\frac{i\pi x_1}{a} \sin\frac{k\pi x_1}{a} \, dx_1 \tag{2.350}$$

has to be calculated. For $i = k$,

$$L_{ik}^* = \int_0^a x_1 \sin^2 \frac{i\pi x_1}{a} \, dx_1 = \frac{a^2}{4} .$$

For $i \neq k$,

$$L_{ik}^* = \int_0^a x_1 \sin\frac{i\pi x_1}{a} \sin\frac{k\pi x_1}{a} dx_1 = \frac{1}{2}\int_0^a x_1 \left[\cos\frac{(i-k)\pi x_1}{a} - \cos\frac{(i+k)\pi x_1}{a} \right] dx_1$$

$$= \frac{1}{2}\left[\frac{a^2}{(i-k)^2\pi^2} \cos(i-k)\pi - \frac{a^2}{(i-k)^2\pi^2} \right]$$

$$- \frac{1}{2}\left[\frac{a^2}{(i+k)^2\pi^2} \cos(i+k)\pi - \frac{a^2}{(i+k)^2\pi^2} \right]$$

For i, k being *both even* or *both odd*,

$$L_{ik}^* = 0.$$

For i *even* (odd) and k *odd* (even),

$$L_{ik}^* = -\frac{4a^2}{\pi^2} \frac{ik}{(i^2 - k^2)^2} .$$

Hence,

$$L_{ik} = -\frac{i^2\pi^2 2}{a^3} L_{ik}^* = \begin{cases} -\dfrac{i^2\pi^2}{2a} & \text{for } i = k, \\ 0 & \text{for } i \neq k, \text{ both even or odd}, \\ \dfrac{8}{a}\dfrac{i^3 k}{(i^2-k^2)^2} & \text{for } i \neq k. \begin{cases} i \text{ even, } k \text{ odd,} \\ i \text{ odd, } k \text{ even.} \end{cases} \end{cases} \tag{2.351}$$

With this result,

$$\int_0^a \int_0^b x_1 \phi_{ij, x_1 x_1} \phi_{kl} dx_1\, dx_2 = -\frac{i^2\pi^2 2}{a^3} \delta_{jl} L_{ik}^*$$

changes into

$$\int_0^a \int_0^b x_1 \phi_{ij, x_1 x_1} \phi_{kl} dx_1\, dx_2 = L_{ik}\, \delta_{jl}. \tag{2.352}$$

Using (2.352) in (2.348) yields

$$\sum_i \sum_j a_{ij} \left\{ \left[\left(-\lambda^* + \lambda_{ij} - g^* \frac{i^2\pi^2}{a} \right) \delta_{ik} - g^* L_{ik} \right] \delta_{jl} \right\} = 0, \tag{2.353}$$

$k, l = 1,2,3, \ldots,$

or

$$\sum_{ij} a_{ij} \rho_{ijkl} = 0, \tag{2.354}$$

where

$$\rho_{ijkl} = \left[\left(-\lambda^* + \lambda_{ij} - g^* \frac{i^2\pi^2}{a} \right) \delta_{ik} - g^* L_{ik} \right] \delta_{jl}. \tag{2.355}$$

The system of algebraic equations (2.354) is homogeneous. In order to ensure nontrivial solutions of the a_{ij},

$$\det \rho_{ijkl} = 0 \tag{2.356}$$

must hold. For an approximate calculation,

$$\begin{vmatrix} \rho_{1111} & \rho_{1211} & \rho_{2111} & \rho_{2211} \\ \rho_{1112} & \rho_{1212} & \rho_{2112} & \rho_{2212} \\ \rho_{1121} & \rho_{1221} & \rho_{2121} & \rho_{2221} \\ \rho_{1122} & \rho_{1222} & \rho_{2122} & \rho_{2222} \end{vmatrix} = 0 \tag{2.357}$$

may be used. By means of (2.355) and (2.351), the following is obtained

$$\rho_{1211} = \rho_{2211} = \rho_{1112} = \rho_{2112} = \rho_{1221} = \rho_{2221} = \rho_{1122} = \rho_{2122} = 0,$$

$$\rho_{1111} = -\lambda^* + \lambda_{11} - g^* \pi^2/2a,$$

$$\rho_{2111} = -64g^*/9a,$$

$$\rho_{1212} = -\lambda^* + \lambda_{12} - g^* \pi^2/2a,$$

$$\rho_{2212} = -64g^*/9a,$$

$$\rho_{1121} = -16g^*/9a,$$

$$\rho_{2121} = -\lambda^* + \lambda_{21} - g^* 4\pi^2/2a,$$

$$\rho_{1222} = -16g^*/9a,$$

$$\rho_{2222} = -\lambda^* + \lambda_{22} - g^* 4\pi^2/2a.$$

Using these values in (2.357) yields

$$\begin{vmatrix} -\lambda^* + \lambda_{11} - g^* \dfrac{\pi^2}{2a} & 0 & -\dfrac{64}{9}\dfrac{g^*}{a} & 0 \\ 0 & -\lambda^* + \lambda_{12} - g^* \dfrac{\pi^2}{2a} & 0 & -\dfrac{64}{9}\dfrac{g^*}{a} \\ -\dfrac{16}{9}\dfrac{g^*}{a} & 0 & -\lambda^* + \lambda_{21} - g^* \dfrac{4\pi^2}{2a} & 0 \\ 0 & -\dfrac{16g^*}{9a} & 0 & -\lambda^* + \lambda_{22} - g^* \dfrac{4\pi^2}{2a} \end{vmatrix} = 0,$$

Interchanging lines and rows appropriately yields

$$\begin{vmatrix} -\lambda^*+\lambda_{11}-g^*\dfrac{\pi^2}{2a} & -\dfrac{64}{9}\dfrac{g^*}{a} & 0 & 0 \\ -\dfrac{16}{9}\dfrac{g^*}{a} & -\lambda^*+\lambda_{21}-g^*\dfrac{4\pi^2}{2a} & 0 & 0 \\ 0 & 0 & -\lambda^*+\lambda_{12}-g^*\dfrac{\pi^2}{2a} & -\dfrac{64}{9}\dfrac{g^*}{a} \\ 0 & 0 & -\dfrac{16}{9}\dfrac{g^*}{a} & -\lambda^*+\lambda_{22}-g^*\dfrac{4\pi^2}{2a} \end{vmatrix} = 0,$$

which is identical to

$$\begin{vmatrix} -\lambda^*+\lambda_{11}-g^*\dfrac{\pi^2}{2a} & -\dfrac{64}{9}\dfrac{g^*}{a} \\ -\dfrac{16}{9}\dfrac{g^*}{a} & -\lambda^*+\lambda_{21}-g^*\dfrac{4\pi^2}{2a} \end{vmatrix} \begin{vmatrix} -\lambda^*+\lambda_{12}-g^*\dfrac{\pi^2}{2a} & -\dfrac{64}{9}\dfrac{g^*}{a} \\ -\dfrac{16}{9}\dfrac{g^*}{a} & -\lambda^*+\lambda_{22}-g^*\dfrac{4\pi^2}{2a} \end{vmatrix} = 0.$$

The eigenvalue curve $F(\lambda^*, g^*) = 0$ is therefore determined by

$$\begin{vmatrix} -\lambda^*+\lambda_{11}-g^*\dfrac{\pi^2}{2a} & -\dfrac{64}{9}\dfrac{g^*}{a} \\ -\dfrac{16}{9}\dfrac{g^*}{a} & -\lambda^*+\lambda_{21}-g^*\dfrac{4\pi^2}{2a} \end{vmatrix} = 0 \qquad (2.358)$$

and

$$\begin{vmatrix} -\lambda^*+\lambda_{12}-g^*\dfrac{\pi^2}{2a} & -\dfrac{64}{9}\dfrac{g^*}{a} \\ -\dfrac{16}{9}\dfrac{g^*}{a} & -\lambda^*+\lambda_{22}-g^*\dfrac{4\pi^2}{2a} \end{vmatrix} = 0 \qquad (2.359)$$

at the same time. By means of (2.343), for $a = b = l$, and using the new variables

$$\xi = \lambda^* l^4 = \frac{\mu\omega^2 l^4}{D}, \quad \eta = g^* l^3 = \frac{g l^3}{D}, \qquad (2.360)$$

equations (2.358), (2.359) yield the eigenvalue equations

$$F(\xi, \eta) = \left(-\xi + 4\pi^4 - \frac{\pi^2}{2}\eta\right)(-\xi + 25\pi^4 - 2\pi^2\eta) - \frac{16 \cdot 64}{81}\eta^2 = 0,$$
$$F(\xi, \eta) = \left(-\xi + 25\pi^2 - \frac{\pi^2}{2}\eta\right)(-\xi + 64\pi^4 - 2\pi^2\eta) - \frac{16 \cdot 64}{81}\eta^2 = 0. \tag{2.361}$$

These analytical expressions correspond to the four branches of the eigenvalue curve in the ξ, η-plane shown in Figure 22.

Let the type of instability be investigated. For this purpose,

$$\frac{\partial F}{\partial \xi} = 0$$

is evaluated. If η turns out to be complex as a consequence of this condition, it is obvious that no real load g exists for which the eigenvalue curve had a horizon-tangent. That means, there is no real load, for which flutter were possible.

Firstly, (2.362) applied to (2.361_1) yields

$$\xi = -\frac{5\pi^2}{4}\eta + \frac{29}{2}\pi^4.$$

Inserting this result in (2.361_1) yields

$$\left(\frac{3}{4}\pi^2\eta - \frac{21}{2}\pi^4\right)^2 + \frac{16 \cdot 64}{81}\eta^2 = 0.$$

Obviously, *this equation cannot be satisfied for real values of* η. Hence, flutter is ruled out.

Secondly, (2.362) is applied to (2.361_2), which leads to

$$\xi = -\frac{5\pi^2}{4}\eta + \frac{89}{2}\pi^2.$$

Using this expression in (2.361_2) yields

$$\left(\frac{3}{4}\pi^2\eta - \frac{39}{2}\pi^4\right)^2 + \frac{16 \cdot 64}{81}\eta^2 = 0.$$

Again, there is no real value η which would satisfy this equation, and flutter is ruled out.

If flutter does not occur, the plate problem in Figure 6 is a *divergence type problem* in spite of its nonconservative follower load g. Therefore, the critical loads causing instability are obtained by intersecting the eigenvalue curve with

Figure 22

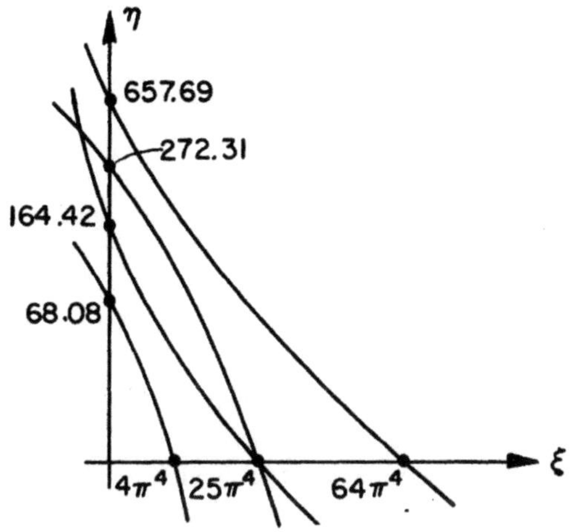

the g-axis. These points of intersection are calculated by setting $\lambda^* = 0$. Due to (2.360) this is equivalent to setting $\xi = 0$ in (2.361). Hence, for $\eta_{crit} = g_{crit}l^3/D$, the two equations

$$\left(4\pi^4 - \frac{\pi^2}{2}\eta\right)(25\pi^4 - 2\pi^2\eta) - \frac{16\cdot 64}{81}\eta^2 = 0,$$
$$\left(25\pi^4 - \frac{\pi^2}{2}\eta\right)(64\pi^4 - 2\pi^2\eta) - \frac{16\cdot 64}{81}\eta^2 = 0 \tag{2.363}$$

are obtained. The first equation yields

$$\eta^2\left(\pi^4 - \frac{64\cdot 16}{81}\right) - \frac{41}{2}\pi^6\eta + 100\pi^8 = 0,$$

$$\eta^2 - 232\eta + 11200 = 0,$$

having the solutions

$$\eta_{crit} = \frac{g_{crit}l^3}{D} = \begin{cases} 69.1 \\ 162.9. \end{cases}$$

The second equation yields

$$\eta^2\left(\pi^4 - \frac{64\cdot 16}{81}\right) - 82\pi^6\eta + 1600\pi^8 = 0,$$

$$\eta^2 - 928\eta + 179200 = 0,$$

having the solutions

$$\eta_{\text{crit}} = \frac{g_{\text{crit}} l^3}{D} = \begin{cases} 275, \\ 653. \end{cases}$$

The actual buckling load of the plate is to be determined by means of the smallest of the four solutions and reads

$$g_{\text{crit}} = 69.1 \frac{D}{l^3}\,.$$

Finally, it may be mentioned that, like for *beam* problems, also for *plate* problems, the corresponding problem involving a conservative, undirectional load g yields a lower bound for g_{crit} of the nonconservative problems, see [30].

Now, let the example in Figure 5 be considered. That is a plate with the edge ratio a/b, freely supported along three edges and free along one of the edges of length l. The plate is subjected to compressive follower forces N which are uniformly distributed along the edges of length b. The follower forces are parallel to the edges of length a. In Section 1.4 the differential equation of this problem was derived. It reads

$$\mu w_{tt} + D\Delta^2 w + N w_{x_1 x_1} = 0. \tag{2.364}$$

The boundary conditions are the following:

$$\begin{aligned} &w = 0 \text{ along } x_1 = 0, && x_2 = 0,\ x_2 = b, \\ &w_{x_1 x_1} + \nu w_{x_2 x_2} = 0 \text{ along} && x_1 = 0,\ x_1 = a, \\ &w_{x_2 x_2} + \nu w_{x_1 x_1} = 0 \text{ along} && x_2 = 0,\ x_2 = b, \\ &w_{x_1 x_1 x_1} + (2-\nu) w_{x_1 x_2 x_2} = 0 \text{ along} && x_1 = a. \end{aligned} \tag{2.365}$$

In agreement with (2.143) and using (2.364) as the integrand, the principle yields

$$\iint\limits_{TB} \{\mu w_{tt} + D\Delta^2 w + N_{w_{x_1 x_1}}\} \delta w dB\, dt = 0. \tag{2.366}$$

The principle is valid in this form only if in

$$w(t, x_1 x_2) = e^{i\omega t} \sum_{i,j} a_{ij} \phi_{ij}(x_1 x_2), \tag{2.367}$$

the coordinate functions ϕ_{ij} satisfy *all* boundary conditions (2.365). By means of (2.367), (2.366) changes into

$$\int\limits_T\int\limits_0^a\int\limits_0^b\left[-\mu\omega^2\sum_{i,j}a_{ij}\phi_{ij}(x_1,x_2)+D\Delta^2\left(\sum_{i,j}a_{ij}\phi_{ij}(x_1,x_2)\right)+N\sum_{i,j}a_{ij}\phi_{ij,x_1x_1}(x_1,x_2)\right]$$

$$\left(\sum_{k,l}\delta a_{kl}\phi_{kl}(x_1,x_2)\right)e^{i\omega t}\,dx_1\,dx_2\,dt=0. \tag{2.368}$$

The eigenfunctions of the free vibration of the unloaded plate may serve as coordinate functions. They satisfy

$$\Delta^2\phi_{ij}=k_{ij}^4\phi_{ij} \tag{2.369}$$

and read

$$\phi_{ij}(x_1,x_2)=\left\{\left[\left(\frac{k_{ij}b}{j\pi}^2\right)-(1-\nu)\right]\sinh\lambda_{2ij}x_1\sin\lambda_{1ij}a\right.$$

$$\left.+\left[\left(\frac{k_{ij}b}{j\pi}\right)^2+(1-\nu)\right]\sinh\lambda_{2ij}a\sin\lambda_{1ij}x_1\right\}\sin\frac{j\pi x_2}{b}\ . \tag{2.370}$$

For the quantities k_{ij}^4, λ_{1ij}, λ_{2ij}, the following expressions hold true

$$k_{ij}^4=\left(\lambda_{2ij}^2-\frac{j^2\pi^2}{b^2}\right)^2, \tag{2.371}$$

$$\lambda_{1ij}=\sqrt{k_{ij}^2-\frac{j^2\pi^2}{b^2}},\quad \lambda_{2ij}=\sqrt{k_{ij}^2+\frac{j^2\pi^2}{b^2}}, \tag{2.372}$$

and they satisfy the condition

$$\sqrt{\left(\frac{k_{ij}b}{j\pi}\right)^2+1}\left[\left(\frac{k_{ij}b}{j\pi}\right)^2-(1-\nu)\right]^2\cosh j\pi\frac{a}{b}\sqrt{\left(\frac{k_{ij}b}{j\pi}\right)^2+1}\cdot\sin j\pi\frac{a}{b}\sqrt{\left(\frac{k_{ij}b}{j\pi}\right)^2-1}$$

$$=\sqrt{\left(\frac{k_{ij}b}{j\pi}\right)^2-1}\left[\left(\frac{k_{ij}b}{j\pi}\right)^2+(1-\nu)\right]^2\sinh j\pi\frac{a}{b}\sqrt{\left(\frac{k_{ij}b}{j\pi}\right)^2+1}\cdot\cos j\pi\frac{a}{b}\sqrt{\left(\frac{k_{ij}b}{j\pi}\right)^2-1}. \tag{2.373}$$

From (2.373), the following numerical values of $(k_{ij}b)^2=(k_{ij}{}^*)^2$ are obtained for $\nu=0.3$ and $a=b$ as displayed in Table 2.5.

Now, it is easily verified that the so chosen ϕ_{ij} contribute to the boundary conditions (2.365) being satisfied by the trial solution (2.367). The last of the conditions (2.365) is specifically satisfied because of relationship (2.373).

Table 2.5

	$j=1$	$j=2$	$j=3$
$i=1$	11.68453	41.19659	90.29402
$i=2$	27.75635	59.06544	108.91840
$i=3$	61.86063	94.48367	145.63800

For the coordinate functions defined by (2.370), the following conditions of orthogonality hold true

$$\int_0^a\int_0^b \phi_{ij}\phi_{kl}\,dx_1\,dx_2 = c_{ijkl}\,\delta_{ik}\,\delta_{jl}, \tag{2.374}$$

and

$$\int_0^a\int_0^b \phi_{ij,x_1x_1}\,\phi_{kl}\,dx_1\,dx_2 = \bar{c}_{ijkl}\,\delta_{jl}, \tag{2.375}$$

where

$$\begin{aligned} c_{ijkl} = \frac{b}{2}\int_0^a &\left\{\left[\left(\frac{k_{ij}b}{j\pi}\right)^2 - (1-\nu)\right]\sinh\lambda_{2ij}x_1\sin\lambda_{1ij}a\right. \\ &\left.+\left[\left(\frac{k_{ij}b}{j\pi}\right)^2 + (1-\nu)\right]\sinh\lambda_{2ij}a\sin\lambda_{1ij}x_1\right\} \\ &\cdot\left\{\left[\left(\frac{k_{kl}b}{l\pi}\right)^2 - (1-\nu)\right]\sinh\lambda_{2kl}x_1\sin\lambda_{1kl}a\right. \\ &\left.+\left[\left(\frac{k_{kl}b}{l\pi}\right)^2 + (1-\nu)\right]\sinh\lambda_{2kl}a\sin\lambda_{1kl}x_1\right\}dx, \end{aligned} \tag{2.376}$$

$$\begin{aligned} \bar{c}_{ijkl} = \frac{b}{2}\int_0^a &\left\{\left[\left(\frac{k_{ij}b}{j\pi}\right)^2 - (1-\nu)\right]\lambda_{2ij}^2\sinh\lambda_{2ij}x_1\sin\lambda_{1ij}a\right. \\ &\left.-\left[\left(\frac{k_{ij}b}{j\pi}\right)^2 + (1-\nu)\right]\lambda_{1ij}^2\sinh\lambda_{2ij}a\sin\lambda_{1ij}x_1\right\} \end{aligned} \tag{2.377}$$

$$\cdot\left\{\left[\left(\frac{k_{kl}b}{l\pi}\right)^2 - (1-\nu)\right] \sinh \lambda_{2kl}x_1 \sin \lambda_{1kl}a\right.$$

$$\left.+\left[\left(\frac{k_{kl}b}{l\pi}\right)^2 + (1-\nu)\right] \sinh \lambda_{2kl}a \sin \lambda_{1kl}x_1\right\} dx.$$

Condition (2.375) is easily verified. The correctness of (2.374) may however be proved through the following calculation.

First, using (1.21) and

$$\begin{aligned} \cos(n, x_1)\, dR &= dx_2, \\ \cos(n, x_2) dR &= -dx_1, \end{aligned} \tag{2.378}$$

which follows from Figure 1, the basic formulae

$$\begin{aligned} &\int_B \left[\frac{\partial^2 u}{\partial x_1^2}\frac{\partial^2 v}{\partial x_2^2} - \frac{\partial^2 u}{\partial x_1 x_2}\frac{\partial^2 v}{\partial x_1 x_2}\right] dB \\ &= \int_R \frac{\partial u}{\partial x_1}\frac{\partial^2 v}{\partial x_2^2}\, dx_2 + \int_R \frac{\partial u}{\partial x_1}\frac{\partial^2 v}{\partial x \partial x_2}\, dx_1 \end{aligned} \tag{2.379}$$

may be derived, which is valid for arbitrary functions u and v. These formulae can be transformed into

$$\begin{aligned} &\int_B \left[\frac{\partial^2 u}{\partial x_1^2}\frac{\partial^2 v}{\partial x_2^2} - \frac{\partial^2 u}{\partial x_1 \partial x_2}\frac{\partial^2 v}{\partial x_1 \partial x_2}\right] dB \\ &= -\int_R \frac{\partial u}{\partial x_1}\frac{\partial^2 v}{\partial_{x_2^2}}\, dx_2 - \int_R u\, \frac{\partial^3 v}{\partial x_1^2 \partial x_2}\, dx_1 \end{aligned} \tag{2.380}$$

by integrating the second term on the right hand side of (2.379) by parts. The boundary term generated by this integration vanishes since the path of integration is closed.

Correspondingly one finds

$$\int_B \left[\frac{\partial^2 u}{\partial x_2^2} \frac{\partial^2 v}{\partial x_1^2} - \frac{\partial^2 u}{\partial x_1 \partial x_2} \frac{\partial^2 v}{\partial x_1 \partial x_2} \right] dB$$
$$= - \int_R \frac{\partial u}{\partial x_2} \frac{\partial^2 v}{\partial x_1^2} dx_1 + \int_R u \frac{\partial^3 v}{\partial x_1 \partial x_2^2} dx_2 . \tag{2.381}$$

Applying (1.21) and (2.378) on the left hand side of (2.380) and (2.381) does not yield unique results. It may also lead to the equivalent relationships

$$\int_B \left[\frac{\partial^2 u}{\partial x_1^2} \frac{\partial^2 v}{\partial x_2^2} - \frac{\partial^2 u}{\partial x_1 \partial x_2} \frac{\partial^2 v}{\partial x_1 \partial x_2} \right] dB$$
$$= - \int_R \frac{\partial v}{\partial x_2} \frac{\partial^2 u}{\partial x_1^2} dx_1 + \int_R v \frac{\partial^3 u}{\partial x_1 \partial x_2^2} dx_2 \tag{2.382}$$

and

$$\int_B \left[\frac{\partial^2 u}{\partial x_2^2} \frac{\partial^2 v}{\partial x_1^2} - \frac{\partial^2 u}{\partial x_1 \partial x_2} \frac{\partial^2 v}{\partial x_1 \partial x_2} \right] dB$$
$$= \int_R \frac{\partial v}{\partial x_1} \frac{\partial^2 u}{\partial x_2^2} dx_2 - \int_R v \frac{\partial^3 u}{\partial x_1^2 \partial x_2} dx_1 . \tag{2.383}$$

Adding (2.380) and (2.381) and then subtracting from this sum (2.382) as well as (2.383) yields after a simple calculation

$$\int_0^b \left\{ \frac{\partial}{\partial x_1} \left[u \frac{\partial^2 v}{\partial x_2^2} - v \frac{\partial^2 u}{\partial x_2^2} \right] \right\} dx_2 = 0, \tag{2.384}$$

which is valid along the *free* edge of the plate, where $dx_1 \equiv 0$. By means of (2.369),

$$(k_{kl}^4 - k_{ij}^4) \int_B \phi_{ij} \phi_{kl} \, dB = \int_B (\phi_{kl} \Delta^2 \phi_{ij} - \phi_{ij} \Delta^2 \phi_{kl}) \, dB \tag{2.385}$$

can be derived. Applying Green's formula

$$\int_B v \Delta^2 u \, dB = \int_B \Delta u \, \Delta v \, dB + \int_R \left[v \frac{\partial \Delta u}{\partial n} - \frac{\partial v}{\partial n} \Delta u \right] dR \tag{2.386}$$

to (2.385) yields

$$(k_{kl}^4 - k_{ij}^4) \int_B \phi_{ij}\phi_{kl}\, dB = \int_R \left[\phi_{kl} \frac{\partial \Delta\phi_{ij}}{\partial n} - \frac{\partial \phi_{kl}}{\partial n} \Delta\phi_{ij} - \phi_{ij} \frac{\partial \Delta\phi_{kl}}{\partial n} + \frac{\partial \phi_{ij}}{\partial n} \Delta\phi_{kl} \right] dR. \tag{2.387}$$

For the edges $x_1 = 0, x_2 = \text{o}$ and $x_2 = b$, the right hand side of (2.387) is always zero, since the ϕ_{ij} and ϕ_{kl} satisfy the boundary conditions. Along the edge $x_1 = a$, $dR \equiv dx_2$, $\partial/\partial n \equiv \partial/\partial x_1$. Hence,

$$(k_{kl}^4 - k_{ij}) \int_B \phi_{ij}\phi_{kl}\, dB = \int_0^b \left[\phi_{kl} \left(\frac{\partial^3 \phi_{ij}}{\partial x_1^3} \frac{\partial^3 \phi_{ij}}{\partial x_1 \partial x_2^2} \right) - \frac{\partial \phi_{kl}}{\partial x_1} \left(\frac{\partial^2 \phi_{ij}}{\partial x_1^2} + \frac{\partial^2 \phi_{ij}}{\partial x_2^2} \right) \right.$$

$$\left. - \phi_{ij} \left(\frac{\partial^3 \phi_{kl}}{\partial x_1^3} + \frac{\partial^3 \phi_{kl}}{\partial x_1 \partial x_2^2} \right) + \frac{\partial \phi_{ij}}{\partial x_1} \left(\frac{\partial^2 \phi_{kl}}{\partial x_1^2} + \frac{\partial^2 \phi_{kl}}{\partial x_2^2} \right) \right] dx_2 \tag{2.388}$$

follows from (2.387). By setting $u = \phi_{kl}, v = \phi_{ij}$ in (2.384), and multiplying this relationship by $(1-\nu)$, one may add this relationship thus obtained to (2.388). Nothing is changed by that procedure, and the result is

$$(k_{kl}^4 - k_{ij}^4) \int_B \phi_{ij}\phi_{kl}\, dB$$

$$= \int_0^b \left[\phi_{kl} \left(\frac{\partial^3 \phi_{ij}}{\partial x_1^3} + \frac{\partial^3 \phi_{ij}}{\partial x_1 \partial x_2^2} \right) - \frac{\partial \phi_{kl}}{\partial x_1} \left(\frac{\partial^2 \phi_{ij}}{\partial x_1^2} + \frac{\partial^2 \phi_{ij}}{\partial x_2^2} \right) \right.$$

$$- \phi_{ij} \left(\frac{\partial^3 \phi_{kl}}{\partial x_1^3} + \frac{\partial^3 \phi_{kl}}{\partial x_1 \partial x_2^2} \right) + \frac{\partial \phi_{ij}}{\partial x_1} \left(\frac{\partial^2 \phi_{kl}}{\partial x_1^2} + \frac{\partial^2 \phi_{kl}}{\partial x_2^2} \right)$$

$$\left. + (1-\nu) \left\{ \frac{\partial}{\partial x_1} \left[\phi_{kl} \frac{\partial^2 \phi_{ij}}{\partial x_2^2} - \phi_{ij} \frac{\partial^2 \phi_{kl}}{\partial x_2^2} \right] \right\} \right] dx_2,$$

or

$$(k_{kl}^4 - k_{ij}^4) \int_B \phi_{ij} \phi_{kl} \, dB \tag{2.389}$$

$$= \int_0^b \left\{ \phi_{kl} \left[\frac{\partial^3 \phi_{ij}}{\partial x_1^3} + (2-\nu) \frac{\partial^3 \phi_{ij}}{\partial x_1 \partial x_2^2} \right] - \frac{\partial \phi_{kl}}{\partial x_1} \left[\frac{\partial^2 \phi_{ij}}{\partial x_1^2} + \nu \frac{\partial^2 \phi_{ij}}{\partial x_2^2} \right] \right.$$

$$\left. - \phi_{ij} \left[\frac{\partial^3 \phi_{kl}}{\partial x_1^3} + (2-\nu) \frac{\partial^3 \phi_{kl}}{\partial x_1 \partial x_2^2} \right] + \frac{\partial \phi_{ij}}{\partial x_1} \left[\frac{\partial^2 \phi_{kl}}{\partial x_1^2} + \nu \frac{\partial^2 \phi_{kl}}{\partial x_2^2} \right] \right\} dx_2,$$

respectively. Along the edge $x_1 = a$, functions ϕ_{ij} and ϕ_{kl} satisfy the second and the fourth of the boundary conditions (2.365). Therefore, it becomes finally obvious that the right hand side of (2.389) is equal to zero, and in fact

$$(k_{kl}^4 - k_{ij}^4) \int_B \phi_{ij} \phi_{kl} \, dB = 0 \tag{2.390}$$

remains of (2.389). If the eigenfunctions ϕ_{ij} and ϕ_{kl} are not the same, $k_{kl}^4 - k_{ij}^4 \neq 0$. Hence, condition (2.390) can only be valid if

$$\int_B \phi_{ij} \phi_{kl} \, dB = 0 \text{ for } \phi_{ij} \neq \phi_{kl}.$$

That proves the orthogonality of the coordinate functions ϕ_{ij}, thus, justifying condition (2.374).

Due to the arbitrariness of the variations δa_{kl} and suppressing

$$\int_T e^{i\omega t} dt,$$

equation (2.368) changes into

$$\int_0^a \int_0^b [-\mu\omega^2 \sum_{i,j} a_{ij} \phi_{ij}(x_1, x_2) + D\Delta^2 \left(\sum_{i,j} a_{ij} \phi_{ij}(x_1, x_2) \right)$$

$$+ N \sum_{i,j} a_{ij} \phi_{ij,x_1 x_1}(x_1, x_2)] \phi_{kl}(x_1, x_2) \, dx_1 \, dx_2 = 0. \tag{2.391}$$

For an approximate calculation, the double series in (2.391) can be assumed to be finite. Consequently, integration and summation can be interchanged. Doing this, taking in addition (2.369), (2.374), (2.375) into account, and using the new notations

$$\lambda^* = \frac{\mu\omega^2 b^4}{D}, N^* = \frac{Nb^2}{D}, k_{ij}b = k_{ij}^*, \frac{c_{ijkl}}{b^2} = c_{ijkl}^*, \tag{2.392}$$

(2.391) yields

$$\sum_{i,j} a_{ij}[(-\lambda^* + (k_{ij}^*)^4)\, c_{ijkl}^*\, \delta_{ik} + N^* \bar{c}_{ijkl}]\, \delta_{jl} = 0, \tag{2.393}$$

$$k, l = 1,2,3, \ldots .$$

This is an algebraic system by which the series coefficients a_{ij} may be determined. By means of

$$\rho_{ijkl} = [(-\lambda^* + (k_{ij}^*)^4)\, c_{ijkl}^*\, \delta_{ik} + N^* \bar{c}_{ijkl}]\, \delta_{jl} \tag{2.394}$$

it changes into

$$\sum_{i,j} a_{ij}\rho_{ijkl} = 0, \qquad k, l = 1,2,3, \ldots . \tag{2.395}$$

It is a homogeneous system. Therefore,

$$\det \rho_{ijkl} = 0 \tag{2.396}$$

must hold. This condition leads to

$$\begin{vmatrix} \rho_{1111} & \rho_{1211} & \rho_{2111} & \rho_{2211} \\ \rho_{1112} & \rho_{1212} & \rho_{2112} & \rho_{2212} \\ \rho_{1121} & \rho_{1221} & \rho_{2121} & \rho_{2221} \\ \rho_{1122} & \rho_{1222} & \rho_{2122} & \rho_{2222} \end{vmatrix} = 0. \tag{2.397}$$

Obviously,

$$\rho_{ijkl} = \begin{cases} 0 \text{ for } j \neq l, \\ \neq 0 \text{ for } j = l. \end{cases} \tag{2.398}$$

Consequently, (2.397) changes into

$$\begin{vmatrix} \rho_{1111} & 0 & \rho_{2111} & 0 \\ 0 & \rho_{1212} & 0 & \rho_{2212} \\ \rho_{1121} & 0 & \rho_{2121} & 0 \\ 0 & \rho_{1222} & 0 & \rho_{2222} \end{vmatrix} = 0,$$

which can be rewritten interchanging lines and rows appropriately to yield

$$\begin{vmatrix} \rho_{1111} & \rho_{21111} & 0 & 0 \\ \rho_{1121} & \rho_{2121} & 0 & 0 \\ 0 & 0 & \rho_{1212} & \rho_{2212} \\ 0 & 0 & \rho_{1222} & \rho_{2222} \end{vmatrix} = 0. \tag{2.399}$$

The eigenvalue equations therefore follow from

$$\begin{vmatrix} \rho_{1111} & \rho_{2111} \\ \rho_{1121} & \rho_{2121} \end{vmatrix} = 0, \quad \begin{vmatrix} \rho_{1212} & \rho_{2212} \\ \rho_{1222} & \rho_{2222} \end{vmatrix} = 0.$$

Taking (2.394) into account, the first determinent reads

$$\begin{vmatrix} (-\lambda^* + (k^*_{11})^4)\, c^*_{1111} + N^* \bar{c}_{1111} & N^* \bar{c}_{2111} \\ N^* \bar{c}_{1121} & (-\lambda^* + (k^*_{21})^4)\, c^*_{2121} + N^* \bar{c}_{2121} \end{vmatrix} = 0$$

yielding

$$\begin{aligned} &\lambda^{*2} c^*_{1111} c^*_{2121} - \lambda^* [(k^*_{21})^4 + (k^*_{11})^4] c^*_{1111} c^*_{2121} \\ &- N^* \lambda^* (c^*_{1111} \bar{c}_{2121} + c^*_{2121} \bar{c}_{1111}) \\ &+ N^* [(k^*_{11})^4 c^*_{1111} \bar{c}_{2121} + (k^*_{21})^4 c^*_{2121} \bar{c}_{1111}] + N^{*2} (\bar{c}_{1111} \bar{c}_{2121} - \bar{c}_{1121} \bar{c}_{2111}) \\ &+ (k^*_{11})^4 (k^*_{21})^4\, c^*_{1111}\, c^*_{2121} = 0. \end{aligned} \tag{2.400}$$

The second determinant reads

$$\begin{vmatrix} (-\lambda^* + (k^*_{12})^4) c^*_{1212} + N^* \bar{c}_{1212} & N^* \bar{c}_{2212} \\ N^* \bar{c}_{1222} & (-\lambda^* + (k^*_{22})^4) c^*_{2222} + N^* \bar{c}_{2222} \end{vmatrix} = 0$$

yielding

$$\begin{aligned} &\lambda^{*2} c^*_{1212} c^*_{2222} - \lambda^* [(k^*_{12})^4 + (k^*_{22})^4] c^*_{1212} c^*_{2222} \\ &- N^* \lambda^* (c^*_{2222} \bar{c}_{1212} + c^*_{1212} \bar{c}_{2222}) \\ &+ N^* [(k^*_{12})^4 c^*_{1212} \bar{c}_{2222} + (k^*_{22})^4 c^*_{2222} \bar{c}_{1212}] \\ &+ N^{*2} (\bar{c}_{1212} \bar{c}_{2222} - \bar{c}_{1222} \bar{c}_{2212}) + (k^*_{12})^4 (k^*_{22})^4 c^*_{1212} c^*_{2222} = 0. \end{aligned} \tag{2.401}$$

The numerical values of (2.376), (2.377) for $a = b$, $\nu = 0.3$ are contained in Table 2.6.

Using the values in Tables 2.5 and 2.6 and the variables

$$\xi = \frac{\lambda^*}{100} = \frac{\mu\omega^2 b^4}{100D}, \qquad \eta = \frac{N^*}{10} = \frac{Nb^2}{10D}, \tag{2.402}$$

the eigenvalue equations (2.400) and (2.401) read as follows:

$$\xi^2 + 1.17\,\xi\eta - 0.177\,\eta^2 - 9.31\,\xi - 1.287\,\eta + 11.09 = 0, \tag{2.403}$$

and

$$\xi^2 + 1.065\,\xi\eta - 0.556\,\eta^2 - 53.24\,\xi - 14.42\,\eta + 624 = 0. \tag{2.404}$$

A graphical representation of these equations is shown in Figure 23. It becomes apparent that the buckling loads are of the *divergence* type. Hence, they can be determined by intersecting the curves (2.403), (2.404) with the η-axis. Hence,

Table 2.6

$c^*_{1111} = 2472.32$	$\bar{c}_{1111} = 134.804\,\pi^2$
$c^*_{1212} = 10728000$	$\bar{c}_{1121} = -2438.9\,\pi^2$
$c^*_{2121} = 177510$	$\bar{c}_{2111} = -19849\,\pi^2$
$c^*_{2222} = 132464000$	$\bar{c}_{1212} = 2411040\,\pi^2$
	$\bar{c}_{2121} = -217651\,\pi^2$
$c_{ijkl} = \dfrac{c^*_{ijkl}}{b^2}$	$\bar{c}_{1222} = -11384100\,\pi^2$
	$\bar{c}_{2212} = -33233100\,\pi^2$
	$\bar{c}_{2222} = -170822000\,\pi^2$

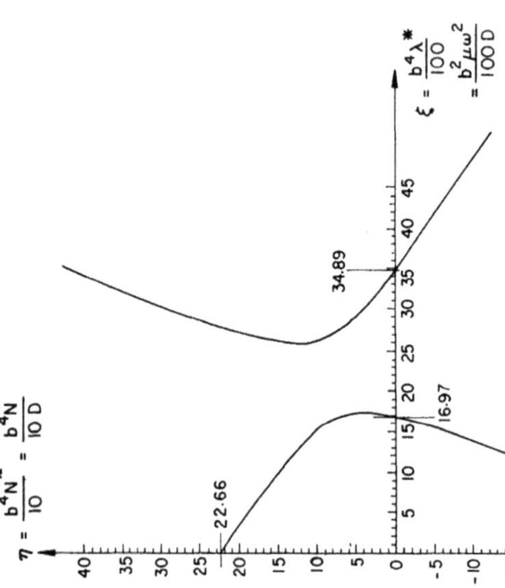

Figure 23b

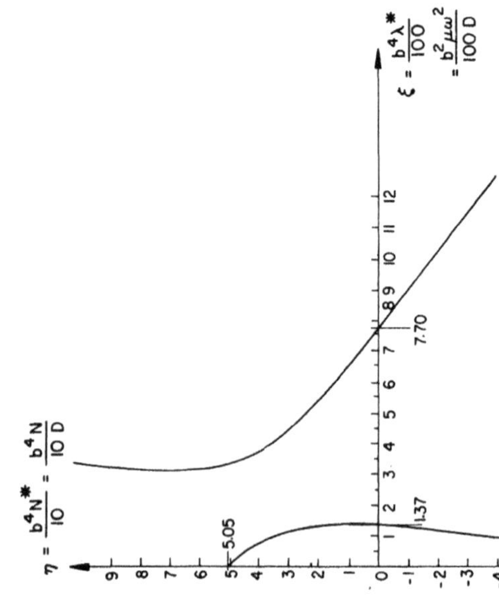

Figure 23a

$\xi = 0$ in (2.403) and (2.404) for a calculation of the buckling loads. This condition leads to

$$-0.177\,\eta^2 - 1.287\,\eta + 11.09 = 0$$

and

$$-0.556\,\eta^2 - 14.42\,\eta + 624 = 0.$$

The first equation has the roots

$$\eta = \begin{cases} 5.072, \\ -12.338, \end{cases} \tag{2.405}$$

and the second equation has the roots

$$\eta = \begin{cases} 22.955, \\ -48.887. \end{cases} \tag{2.406}$$

By means of (2.402), the positive values in (2.405), (2.406), respectively, yield the buckling loads

$$N_{\text{crit},1} = 5.072\,\frac{D}{b^2}\cdot 10 = 50.72\,\frac{D}{b^2}$$

and

$$N_{\text{crit},2} = 22.955\,\frac{D}{b^2}\cdot 10 = 229.55\,\frac{D}{b^2}\,.$$

3

The method of Galerkin

3.1 Extended equations of Galerkin. Conservative problems

As already mentioned repeatedly, this case is characterized by $Q[w] = 0, P[w] = 0$. The fundamental variational principle (1.19) becomes therefore the principle of Hamilton and reads in this form:

$$\int_T \left[\int_B \left\{ \frac{d}{dt} \left(\frac{\partial \mathscr{T}}{\partial w_t} \right) \delta w + \delta \mathscr{U} \right\} dB \right] dt = 0. \tag{3.1}$$

Restricting oneself to one-dimensional problems means that one has to deal with

$$\mathscr{T} = \frac{1}{2} \mu \omega^2, \tag{1.55}$$

$$\mathscr{U} = \frac{1}{2} \sum_{k=0}^{2} h_k(x) \left[w^{(k)}(x, t) \right]^2, \tag{2.4}$$

$$w(x, t) = f(t) y(x), \tag{1.100}$$

$$c f_{tt} = Tf = -\lambda f. \tag{1.101}$$

Using these expressions in (3.1), calculating $\delta \mathscr{U}$ and suppressing

$$\int_T f^2 \, dt,$$

makes (3.1) change into the already known relationship

$$\int_B \left\{ -\lambda y \, \delta y + \sum_{k=0}^{2} h_k y^{(k)} \delta y^{(k)} \right\} dB = 0. \tag{2.10}$$

By means of

$$y^{(k)} = \sum_i a_i \phi_i^{(k)}, \qquad k = 0,1,2, \tag{2.11}$$

equation (2.10) yields the *equations of Ritz*

$$\sum_i a_i \left\{ -\lambda\delta_{ij} + \int_B \sum_{k=0}^{2} h_k \phi_i^{(k)} \phi_j^{(k)} dB \right\} = 0, \qquad j = 1,2,3, \ldots . \tag{2.14}$$

The derived variational principle

$$\int_T \left[\int_B \left\{ \frac{d}{dt}\left(\frac{\partial \mathcal{T}}{\partial w_t}\right) + \frac{\delta U}{\delta w} \right\} \delta w \, dB + \int_R W(w, \delta w)\, dR \right] dt = 0 \tag{3.2}$$

is equivalent to the fundamental principle. It follows from (1.46) by the vanishing of $Q[w] = Q_t[w] + Q_x[w]$ and $P[w]$. The expression $W(w, \delta w)$ is defined by (1.43). Corresponding to the preceding procedure,

$$\int_B \left\{ -\lambda y + \sum_{k=0}^{2} (-1)^{(k)} [h_k y^{(k)}]^{(k)} \right\} \delta y \, dB + \int_R W(y, \delta y)\, dR = 0 \tag{3.3}$$

is obtained from (3.2). This is achieved by means of (1.55), (2.4), (1.100), (1.101), by using (1.42) in the form

$$\frac{\delta U}{\delta w} = \frac{\partial \mathcal{U}}{\partial w} - \frac{\partial}{\partial x}\left(\frac{\partial \mathcal{U}}{\partial w_x}\right) + \frac{\partial^2}{\partial x^2}\left(\frac{\partial \mathcal{U}}{\partial w_{xx}}\right), \tag{3.4}$$

and by suppressing $\int_T f^2 dt$. According to (1.43), and taking the fact into account that we are dealing with one-dimensional problems,

$$W(y, \delta y) = \left(\frac{\partial \mathcal{U}}{\partial y'} - \frac{\partial}{\partial x}\frac{\partial \mathcal{U}}{\partial y''}\right)\delta y + \frac{\partial \mathcal{U}}{\partial y''}\,\delta y'.$$

This expression becomes

$$W(y, \delta y) = \{h_1(x) y' - [h_2(x) y'']'\}\delta y + h_2(x)''\,\delta y' \tag{3.5}$$

by virtue of (2.4). In addition, the integral

$$\int_R W(y, \delta y)\, dR$$

degenerates into the boundary term $[W(y, \delta y)]_R$ if the problem is onedimensional. Hence, (3.3) changes into

$$\int_B \left\{ -\lambda y + \sum_{k=0}^{2} (-1)^k [h_k y^{(k)}]^{(k)} \right\} \delta y \, dB$$

$$+ [\{h_1(x) y' - [h_2(x) y'']'\} \delta y + h_2(x) y'' \delta y']_R = 0. \tag{3.6}$$

By means of (2.11),

$$\sum_i a_i \left\{ -\lambda \delta_{ij} + \int_B \sum_{k=0}^{2} (-1)^k [h_k \phi_i^{(k)}]^{(k)} \phi_j \, dB \right.$$

$$\left. + [\{h_1(x) \phi_i' - [h_2(x) \phi_i'']'\} \phi_j + h_2(x) \phi_i'' \phi_j']_R \right\} = 0 \tag{3.7}$$

$$k = 1,2,3, \ldots$$

follows from (3.6). In the most general case it may be assumed that the coordinate functions used in the equations of Ritz (2.14) are admissible functions, i.e., that they satisfy the *geometric* boundary conditions only. Then,

$\phi_i(0) = \phi_i(l) = 0$	for the pinned-pinned rod,
$\phi_i(0) = \phi_i'(0) = \phi_i(l) = \phi_i'(l) = 0$	for the clamped-clamped rod,
$\phi_i(0) = \phi_i'(0) = \phi_i(l) = 0$	for the clamped-pinned rod,
$\phi_i(0) = \phi_i'(0) = 0$	for the clamped-free rod.

As a consequence of these boundary conditions, and of the assumption that $h_1(l) = 0$, $h_2 = \alpha = \text{const.}$, equation (3.7) assumes the form

$$\sum_i a_i \left\{ -\lambda \delta_{ij} + \int_B \sum_{k=0}^{2} (-1)^k [h_k \phi_i^{(k)}]^{(k)} \phi_j \, dB + [\ldots]_R \right\} = 0, \tag{3.8}$$

$$k = 1,2,3, \ldots,$$

where

$$\begin{aligned}
&[\dots]_R = [\alpha\phi_i''\phi_j']_0^l && \text{for the pinned-pinned rod,}\\
&[\dots]_R = 0 && \text{for the clamped-clamped rod,}\\
&[\dots]_R = \alpha\phi_i''(l)\phi_j'(l) && \text{for the clamped-pinned rod,}\\
&[\dots]_R = -\alpha\phi_i'''(l)\phi_j(l) + \alpha\phi_i''(l)\phi_j'(l) && \text{for the clamped-free rod.}
\end{aligned} \tag{3.9}$$

Relationships (3.8), (3.9) represent the system of the "*extended*" *equations of Galerkin*. For multiple dimensional problems, corresponding equations of relatively higher complexity can be found.

The extended equations of Galerkin (3.8), (3.9) are completely analogous to the equations of Ritz (2.14). They follow from those by means of mathematical transformations. Therefore, they do not have any importance of their own. Also, the convergence of the extended equations of Galerkin has not to be investigated separately. In Section 1.2 it has been shown that the equations of Ritz converge using coordinate functions satisfying the geometric boundary conditions only. But the extended equations of Galerkin are only a special form of the equations of Ritz and can be transformed back into those. Hence, the convergence of the extended equations of Galerkin is ensured since the convergence of the equations of Ritz has been proved.

3.2 The classical case. Conservative problems

In the classical case, which is the one for which Galerkin did once introduce his method, it is assumed that the coordinate functions ϕ_i satisfy *all* the boundary conditions of the problem concerned. The equations of Ritz (2.14) are not affected at all by this assumption in their formal appearance. On the other hand,

$$\int_R W(\phi_i, \phi_j)\, dR \equiv 0, \qquad [W(\phi_i, \phi_j)]_R \equiv 0, \text{ respectively,}$$

for these specific coordinate functions. Therefore, the boundary term $[\dots]_R$, given by (3.9), vanishes completely in the extended equations of Galerkin (3.8). Hence, from (3.8) follows in this case

$$\sum_i a_i \left\{ -\lambda\delta_{ij} + \int_B \sum_{k=0}^{2} (-1)^k [h_k \phi_i^{(k)}]^{(k)} \phi_j \, dB \right\} = 0. \tag{3.10}$$

These are the *equations of Galerkin in its classical form*. For these, the remarks made in Section 3.1 apply also. They are analogous to the equations of Ritz

(2.14), from which they can be obtained or into which they can easily be transformed by means of formal operations. Therefore, they do not have an independent significance. Their convergence can easily be proved: they may be changed into the corresponding equations of Ritz, and they converge therefore if those do converge. In Section 2.1 it has already been mentioned that the equations of Ritz converge if they are formed using coordinate functions satisfying all boundary conditions. Consequently, the convergence of the classical equations of Galerkin for conservative problems is ensured.

3.3 Generalized equations of Ritz. Extended equations of Galerkin. Nonconservative problems

If the problems under consideration are nonconservative ones, the fundamental principle (1.19) has full effect on the solution of the problem. The principle shall be used here and in the following in a slightly restricted form in so far as friction will not be taken into account. That means $Q_t\,[w] \equiv 0$. Therefore, the principle reads

$$\int_T \left[\int_B \left\{ \frac{d}{dt}\left(\frac{\partial \mathscr{T}}{\partial w_t}\right) \delta w + \delta \mathscr{U} - Q_x[w]\delta w \right\} dB - \int_R P[w]\delta w\, dR \right] dt = 0. \tag{3.11}$$

Using it for one-dimensional problems, and taking "follower" forces as nonconservative loads into account, one has to work in (3.11) with the following quantities:

$$\mathscr{T} = \frac{1}{2}\,\mu w_t^2, \tag{1.55}$$

$$\mathscr{U} = \frac{1}{2} \sum_{k=0}^{2} h_k(x)\,[w^{(k)}(x,t)]^2, \tag{2.4}$$

$$Q_x[w] = -g w_x, \tag{2.123}$$

$$P[w] = -P w_x. \tag{2.124}$$

Moreover, the integral

$$\int_R P[w]\,\delta w\, dR$$

changes into the boundary term

$$[-P w_x\, \delta w]_R \tag{3.12}$$

in the one-dimensional case. Since at least one end of the rod is supported, it may be the end at which $x = 0$, the variation $\delta w = 0$ there. Hence, from (3.12) remains

$$-Pw_x(l)\,\delta w(l). \tag{3.13}$$

For solutions of the form

$$w(x, t) = f(t)\,y(x), \tag{1.100}$$

$$cf_{tt} = Tf = -\lambda f, \tag{1.101}$$

so that by virtue of this, and after suppressing $\int_T f^2\,dt$, the principle (3.11) changes into

$$\int_0^l \left\{ -\lambda y\,\delta y + \sum_{k=0}^{2} h_k y^{(k)} \delta y^{(k)} + gy'\delta y \right\} dx + Py'(l)\,\delta y(l) = 0. \tag{3.14}$$

By means of $y^{(k)} = \sum_i a_i \phi_i^{(k)}$, $k = 0,1,2$, (3.14) yields

$$\sum_i a_i \left\{ -\lambda \delta_{ij} + \int_0^l \left[\sum_{k=0}^{2} h_k \phi_i^{(k)} \phi_j^{(k)} + g\phi_i'\phi_j \right] dx + P\phi_i'(l)\,\phi_j(l) \right\} = 0. \tag{3.15}$$

These equations correspond to (2.132), (2.206), respectively, and may be called *generalized equations*. They differ from the usual equations of Ritz (2.14) in so far as they contain additional terms taking account of the work of the nonconservative forces.

The derived variational principle, which, according to (1.46) and for $Q_t[w] = 0$ reads

$$\int_T \left[\int_B \left\{ \frac{d}{dt}\left(\frac{\partial \mathscr{T}}{\partial w_t}\right) + \frac{\delta U}{\delta w} - Q_x[w]\,\delta w \right\} dB \right.$$

$$\left. + \int_R \left\{ W(w, \delta w) - P[w]\,\delta w \right\} dR \right] dt = 0, \tag{3.16}$$

is equivalent to the fundamental principle. According to the preceding deliberations, (3.16) changes into

$$\int_B \{-\lambda y + \sum_{k=0}^{2} (-1)^k [h_k y^{(k)}]^{(k)} + gy'\}\,\delta y\,dB + [W(y, \delta y) + Py'\delta y]_R = 0. \tag{3.17}$$

This is achieved by using (1.55), (2.4), (2.123), (2.124), (3.3), (3.12), (1.100), (1.101), by suppressing $\int_T f^2\, dt$ and by taking into account that in the one-dimensional case the integral $\int_R W(w, \delta w)\, dR$ changes into the boundary term $[W(w, \delta w)]_R$. It has already been shown that for $[W(y, \delta y)]_R$ the expression (3.5) applies, and that due to the support conditions of the rod, $[Py'\delta y]_R = Py'(l)\,\delta y(l)$. Hence (3.17) can be replaced by

$$\int_B \left\{-\lambda y + \sum_{k=0}^{2} (-1)^k [h_k y^{(k)}]^{(k)} + gy'\right\} \delta y\, dB \tag{3.18}$$

$$+\Big[\{h_1(x)y' - [h_2(x)y'']'\}\,\delta y + h_2(x)y''\,\delta y'\Big]_R + Py'(l)\,\delta y(l) = 0.$$

By means of (2.11) and by virtue of $B = 0$ or l, $dB = dx$, (3.18) yields the system of equations

$$\sum_i a_i \left\{-\lambda\delta_{ij} + \int_0^l \left[\sum_{k=0}^{2} (-1)^k [h_k \phi_i^{(k)}]^{(k)} + g\phi_i'\right] \phi_j\, dx \right. \tag{3.19}$$

$$\left. +\Big[\{h_1(x)\phi_i' - [h_2(x)\phi_i'']'\}\phi_j + h_2(x)\phi_i''\,\phi_j'\Big]_R + P\phi_i'(l)\,\phi_j(l)\right\} = 0,$$

$j = 1,2,3, \ldots$.

In the most general case it can be assumed that the coordinate functions ϕ_i are admissible functions which satisfy only the *geometric* boundary conditions. Then, the terms

$$\Big[\{h_1(x)\phi_i' - [h_2(x)\phi_i'']'\}\phi_j + h_2(x)\phi_i''\,\phi_j'\Big]_R + P\phi_i'(l)\,\phi_j(l)$$

become specifically equal to

$[\alpha\phi_i''\,\phi_j']_0^l$	for the pinned-pinned rod,
zero	for the clamped-clamped rod,
$\alpha\phi_i''(l)\,\phi_j'(l)$	for the clamped-pinned rod,
$-\alpha\phi_i'''(l)\,\phi_j(l) + \alpha\phi_i''(l)\,\phi_j'(l) + P\phi_i'(l)\,\phi_j(l)$	for the clamped-free rod.

Moreover, it has been assumed that $h_2(x) \equiv \alpha = \text{const.}$

System (3.19) together with conditions (3.20) represent the *"extended" equations of Galerkin*. They differ from those which are valid in the conservative case in so far as additional terms are present in comparison to (3.8), which correspond to the work done by the nonconservative forces, and that (3.20) has been substituted for (3.9)

The convergence of such extended equations has already been investigated in Section 2.2 with an answer in the affirmative. Therefore, there is no need to repeat such discussions here. Since due to the equivalence of the fundamental and the derived variational principle only a formal difference exists between the generalized equations of Ritz and the extended equations of Galerkin in the nonconservative case, it is sufficient to prove the convergence of the one system, thus implying immediately the convergence of the other system. These ideas have already been used in Section 2.2, where the convergence of the (generalized) equations of Ritz were proved by proving the convergence of the (extended) equations of Galerkin which are equivalent to them.

3.4 Classical equations of Galerkin. Nonconservative problems

In the classical case, coordinate functions ϕ_i are being used which satisfy *all* boundary conditions of the posed problem. The generalized equations of Ritz (3.15) do not change following this assumption. However, the extended equations of Galerkin (3.19) are affected. The boundary term $[\ldots]_R$ vanishes immediately and the term $P\phi_i'(l)\phi_j(l)$ is different from zero only for the clamped-free rod. Hence,

$$\sum_i a_i \left\{ -\lambda\delta_{ij} + \int_0^l \left[\sum_{k=0}^{2} (-1)^k [h_k \phi_i^{(k)}]^{(k)} + g\phi_i' \right] \phi_j \, dx + P\phi_i'(l)\phi_j(l) \right\} = 0,$$

$$j = 1,2,3,\ldots \tag{3.21}$$

remains as the system of classical equations of Galerkin for the clamped-free rod and

$$\sum_i a_i \left\{ -\lambda\delta_{ij} + \int_0^l \left[\sum_{k=0}^{2} (-1)^k [h_k \phi_i^{(k)}]^{(k)} + g\phi_i' \right] \phi_j \, dx \right\} = 0 \tag{3.22}$$

for all other, completely supported rods.

As mentioned in Section 2.2, the classical equations of Galerkin converge for comparison functions (which satisfy all boundary conditions) even better than for "admissible" coordinate functions (which satisfy the geometric boundary conditions only). But the classical equations of Galerkin do converge for admissible coordinate functions, as has been shown in Section 2.2. Hence, the classical equations of Galerkin do converge in the nonconservative case together with the corresponding and equivalent generalized equations of Ritz.

3.5 Classical equations of Galerkin. Quasilinear problems

Consider the nonlinear boundary value problem

$$D[x(t)] + cx = f[t, x(t)], \tag{3.23}$$

$$(Gx)_R = 0. \tag{3.24}$$

The differential operator $D[x(t)]$ be linear, selfadjoint with respect to (3.24), semi-definite, and of order $m \geqq 4$. The nonlinear function $f[t, x(t)]$ be continuous, bounded, Lipschitz bounded in the interval I of t defined by R. Let I be given as $a \leqq t \leqq b$. The coefficient c in (3.23) can be assumed to be always positive without loss of generality.

The problem (3.23), (3.24) is a so-called quasi-linear problem according to its specific structure. It is closely related to the linear "auxiliary problem"

$$D[x(t)] + cx = -\kappa x, \tag{3.25}$$

$$(Gx)_R = 0, \tag{3.24}$$

which may be rewritten to yield

$$D[x(t)] = -(c + \kappa)\, x = -\lambda x, \tag{3.26}$$

$$(Gx)_R = 0. \tag{3.24}$$

Problem (3.26), (3.24) may have the eigenfunctions $x_k(t)$ and the eigenvalues λ_k, $k = 1,2,3, \ldots$. Since $D[x(t)]$ is semi-definite, all λ_k have the same sign, however zero may be an eigenvalue. Problem (3.25), (3.24) has the same eigenfunctions $x_k(t)$, but the eigenvalues $\kappa_k = \lambda_k - c$. According to the magnitude of c a finite number of eigenvalues κ_k may have the opposite sign as all the other eigenvalues. However, always $c \neq \lambda_k$, so that the auxiliary problem (3.25), (3.24) can never have the eigenvalue zero.

Due to the selfadjointness of $D[x(t)]$, the eigenfunctions x_k are orthogonal in the actual and in a generalized sense. In addition they may be normalized. Then,

$$\int_a^b x_i x_k \, dt = \delta_{ik}, \tag{3.27}$$

$$\int_a^b x_i D x_k \, dt = -\lambda_i \delta_{ik}. \tag{3.28}$$

The method of Galerkin in its classical form consists in using for the solution $x(t)$ of problem (3.23), (3.24) a series expansion involving coordinate functions which satisfy *all* boundary conditions (3.24). This is certainly the case if the eigenvalue functions of the linear auxiliary problem are used as coordinate functions. In that case the expression

$$x(t) = \sum_{k=1}^{\infty} a_k x_k(t) \tag{3.29}$$

is convergent, since $x(t)$ is with respect to problem (3.25), (3.24) an admissible function. The main objective of the method is to determine the coefficients a_k in series (3.29). For this purpose, (3.29) is inserted into (3.23). Then, both sides in (3.23) are multiplied by $x_i(t)$ and integrated over the interval I. The result is

$$\begin{aligned} &\int_a^b \left\{ D\left[\sum_k a_k x_k(\xi) \right] + c\left[\sum_k a_k x_k(\xi) \right] \right\} x_i(\xi) \, d\xi \\ &= \int_a^b f\left[\xi, \sum_k a_k x_k(\xi) \right] x_i(\xi) \, d\xi. \end{aligned} \tag{3.30}$$

Using in (3.30) the fact that D is linear, and using the relationships (3.27), (3.28) yields the equations of Galerkin in the form

$$(-\lambda_i + c) a_i = \int_a^b f\left[\xi, \sum_k a_k x_k(\xi) \right] x_i(\xi) \, d\xi.$$

Hence, for a calculation of the coefficients a_k in (3.29), one may apply the infinite, nonlinear system of equations

$$a_i = -\frac{1}{\lambda_i - c} \int_a^b f\left[\xi, \sum_{k=1}^{\infty} a_k x_k(\xi) \right] x_i(\xi) \, d\xi. \tag{3.31}$$

The unknowns a_k of this system may be evaluated by means of the reduction method, and will be obtained in the course of successive approximations. One arrives in this way at the approximate solution

$$x^{(n)} = \sum_{k=1}^{n} a_k^{(n)} x_k \tag{3.32}$$

instead of (3.29). This procedure is justified, if $x^{(n)}$ approaches the exact solution (3.23), (3.24) to any required degree of accuracy. As will be shown, this is the case if $f[\xi, x(\xi)]$ satisfies certain sufficient conditions.

As the linear auxiliary problem (3.25), (3.24) does not have the eigenvalue zero, there exists a Green function $G(t, \xi)$ with respect to (3.25), (3.24) which is continuous and symmetric. Furthermore, it can be assumed that $G(t, \xi)$ is bounded, i.e.,

$$|G(t, \xi)|_{\max} < C_1 . \tag{3.33}$$

Then, the integral equation

$$\phi(t) = -\kappa \int_a^b G(t, \xi)\phi(\xi)\, d\xi \tag{3.34}$$

can be associated with the auxiliary problem (3.25), (3.24) as its equivalent counterpart. The eigenfunctions of this integral equation are the eigenfunctions x_k and its eigenvalues are the eigenvalues κ_k of (3.25), (3.24). Since, as already stated, the κ_k have all the same sign with the exception of a finite number of them, the theorem of Mercer applies, according to which

$$G(t, \xi) = \sum_{k=1}^{\infty} \frac{x_k(t) x_k(\xi)}{\lambda_k - c} . \tag{3.35}$$

The series on the right hand side in (3.35) converges absolutely and uniformly.

Knowing Green's function (3.35) attached to the differential operator $D[x] + cx$ in the presence of boundary conditions (3.24), the integral equation

$$\begin{aligned} x(t) &= -\int_a^b G(t, \xi) f[\xi, x(\xi)]\, d\xi \\ &= -\int_a^b \sum_{k=1}^{\infty} \frac{x_k(t) x_k(\xi)}{\lambda_k - c} f[\xi, x(\xi)]\, d\xi \end{aligned} \tag{3.36}$$

can be used to substitute the original problem (3.23), (3.24). Inserting in (3.36) for x the series (3.29) and comparing coefficients of $x_k(t)$ on both sides of (3.36), one arrives at

$$a_i = -\frac{1}{\lambda_i - c} \int_a^b f\left[\xi, \sum_{k=1}^{\infty} a_k x_k(\xi)\right] x_i(\xi)\, d\xi, \tag{3.31}$$

which is a system of equations introduced already earlier. One recognizes that solving the infinite, nonlinear system (3.31) is completely equivalent to solving the nonlinear integral equation (3.36). This integral equation is similar to a Hammerstein equation [39]. The difference consists in the fact that in (3.36) the kernel is not positive definite as it is the case for the kernel in a Hammerstein equation.

The practical approach to solving (3.31) may consist in substituting the infinite system by a finite one, i.e., by

$$a_i^{(n)} = -\frac{1}{\lambda_i - c} \int_a^b f\left[\xi, \sum_{k=1}^{n} a_k^{(n)} x_k(\xi)\right] x_i(\xi)\, d\xi, \tag{3.37}$$

and in trying to approximate the solution successively by letting n increase and hoping that $a_k^{(n)} \to a_k$ for $n \to \infty$.

Like for (3.31), an equivalent integral equation can be set up for (3.37) by multiplying both sides by $x_i(t)$ and then adding over i. Since one has to deal with a finite sum in (3.37), the order of summation and integration can be interchanged. The result is

$$\sum_{i=1}^{n} a_i^{(n)} x_i(t) = -\int_a^b \sum_{i=1}^{n} \frac{x_i(t)\, x_i(\xi)}{\lambda_i - c} f\left[\xi, \sum_{k=1}^{n} a_k^{(n)} x_k(\xi)\right] d\xi.$$

Using

$$x^{(n)}(t) = \sum_{k=1}^{n} a_k^{(n)} x_k(t), \quad G^{(n)}(t, \xi) = \sum_{k=1}^{n} \frac{x_k(t)\, x_k(\xi)}{\lambda_k - c}, \tag{3.38}$$

$$x^{(n)}(t) = -\int_a^b G^{(n)}(t, \xi) f[\xi, x^{(n)}(\xi)]\, d\xi \tag{3.39}$$

follows from the previous equations thus showing that it is the approximative version of integral equation (3.36).

The conclusion is that the successive approximation of the infinite system of Galerkin's equations (3.31) by means of the reduction steps (3.37) corresponds completely to the successive approximation of integral equation (3.36) by means of (3.39). Therefore, the investigation of the convergence of the approximative procedure (3.37) can be replaced by investigating instead the convergence of the approximative procedure (3.39). This is a simpler matter since integral equations are more manageable.

Consider the set Ω of all continuous, bounded and therefore quadratically integrable functions. By virtue of the boundedness of $f[\xi, x(\xi)]$ and the continuity and boundedness of $G(t, \xi)$, also the fixed point $x_0 = Tx_0$ belongs to Ω, if it exists. Now consider the subset $\Omega_0 \subset \Omega$, which is defined by

$$x = \{x \epsilon \Omega_0 : |x - x_0| < \delta\}, \quad \delta \geqq 0, \; x_0 \epsilon \Omega_0. \tag{3.40}$$

This is the set of all continuous, bounded and therefore quadratically integrable functions *in the neighbourhood of* x_0.

Equation (3.36) can be represented as $x = Tx$ by means of the integral operator

$$Tx = -\int_a^b G(t, \xi) f[\xi, x(\xi)]\, d\xi. \tag{3.41}$$

It was already assumed that function f be bounded. In addition, it may be Lipschitz-bounded on Ω_0 so that

$$|f(x) - f(\bar{x})| \leqq \alpha\, |x - \bar{x}|. \tag{3.42}$$

By virtue of this property, and because of the continuity and boundedness of G in the interval $I \times I$, the operator T maps the set Ω_0 compactly into itself. This can be seen as follows: Let $y = Tx$ be the image produced by T. Then,

$$|y| \leqq (b-a)\, |G|_{max}\, |f|_{max},$$

i.e., the set of images y in I is uniformly bounded. Moreover,

$$|y(t_1) - y(t_2)| \leqq (b-a)\, |G(t_1, \xi) - G(t_2, \xi)|_{max}\, |f|_{max}$$

in I. Due to the continuity of G, the set of images y is also uniformly continuous. According to the theorem of Arzela, the set of images y produced by the operator T is therefore compact. Finally, the mapping does not lead outside of Ω_0, because due to (3.42),

$$|y - x_0| \leqq (b-a)\, |G|_{max}\, \alpha\, |x - x_0|.$$

Under the assumption

$$\alpha < \frac{1}{(b-a)\, G_{max}}, \tag{3.43}$$

and taking (3.40) into account, it follows that

$$|y - x_0| < \delta,$$

i.e., $y \epsilon \Omega_0$.

Since the operator T defined by (3.41) maps the set Ω_0 compactly into itself, it possesses at least one fixed point according to the theorem of Schauder [37]. Hence, the equation $x = Tx$ and therefore the integral equation (3.36) has at least one solution. Due to the correspondence of (3.36) with (3.31) also the infinite system of equations (3.31) has at least one solution for the coefficients a_i, and the representation of the solution $x(t)$ of problem (3.23), (3.24) by means of (3.29) is meaningful.

Now, it shall be shown that the solution x_0 of (3.36) can be obtained using a process of successive approximation. For this purpose, the equation of approximation (3.39) may be slightly modified and rewritten as

$$x^{(n+1)}(t) = \int_a^b G^{(n)}(t, \xi) f[\xi, x^{(n)}(\xi)]\, d\xi. \tag{3.44}$$

From (3.36) and (3.44), and introducing the abbreviation

$$f[\xi, x^n(\xi)] = f^{(n)}, \tag{3.45}$$

the estimate

$$|x - x^{(n+1)}| \leqq \int_a^b |Gf - Gf^{(n)} + Gf^{(n)} - G^{(n)} f^{(n)}|\, d\xi$$

is obtained, which can be transformed into

$$|x - x^{(n+1)}| < (b-a) C_1 \alpha |x - x^{(n)}| + (b-a) |G - G^{(n)}|_{\max} |f^{(n)}|_{\max} \tag{3.46}$$

using (3.33) and (3.42). Because of the convergence of (3.35), $G^{(n)} \to G$ for $n \to \infty$, and

$$|G - G^{(n)}|_{\max} = r^{(n)} \to 0 \text{ for } n \to \infty$$

holds true. Inequality (3.46) yields after $k-1$ more iteration steps

$$|x - x^{(n+k)}| < [(b-a) C_1 \alpha]^k |x - x^{(n)}| + \frac{1 - [(b-a) C_1 \alpha]^k}{1 - [(b-a) C_1 \alpha]} (b-a) M r^{(n)}, \tag{3.47}$$

where M is the maximum of all $|f^{(p)}|_{\max}$, $p = n, n+1, \ldots, n+k-1$. Due to (3.43) also

$$(b-a)\,C_1\,\alpha < 1. \tag{3.48}$$

Hence, (3.47) yields by virtue of

$$[(b-a)\,C_1\,\alpha]^k \to 0 \text{ for } k \to \infty,$$

$$r^{(n)} \to 0 \text{ for } n \to \infty$$

also $|x - x^{(n+k)}| \to 0$ for $n, k \to \infty$. The iterative process (3.44), and hence the method of Galerkin which is equivalent to (3.44), converges. The process converges toward the fixed point x_0 in Ω_0. Because if the convergence of (3.44) were toward $\tilde{x} \epsilon \Omega_0$, then, due to (3.42),

$$x_0(t) - \int_a^b G(t,\xi) f[\xi, x_0(\xi)]\, d\xi,$$

and

$$\tilde{x}(t) = - \int_a^b G(t,\xi) f[\xi, \tilde{x}(\xi)]\, d\xi,$$

also

$$|x_0 - \tilde{x}| < (b-a)\,C_1\,\alpha\, |x_0 - \tilde{x}|.$$

This leads to a contradiction, if $x_0 = \tilde{x}$ were not correct, in the light of (3.48).

The convergence of the method of Galerkin can still be proved even if the function f involves derivatives of x. In order to keep the following deliberations simple, it may be assumed that f contains only the first derivative $\dot{x}$. Then, the problem under consideration reads

$$D[x(t)] + cx = f[t, x(t), \dot{x}(t)], \tag{3.49}$$

$$(Gx)_R = 0. \tag{3.24}$$

Due to the very nature of the linear auxiliary problem (3.25), (3.24), the Green function is differentiable with respect to t, and the series expansions

$$\frac{\partial^k G(t,\xi)}{\partial t^k} = \sum_{i=1}^{\infty} \frac{x_i^{(k)}(t)\, x_i(\xi)}{\lambda_i - c}, \qquad k \leq \frac{m}{2} - 1 \tag{3.50}$$

are convergent [40]. Moreover,

$$\left|\frac{\partial^k G(t,\xi)}{\partial t^k}\right|_{\max} < C_2 \text{ for } k = 1,2,3,\ldots,m. \tag{3.51}$$

This time,

$$x(t) = -\int_a^b G(t,\xi) f[\xi, x(\xi), \dot{x}(\xi)]\, d\xi \tag{3.52}$$

instead of (3.36) and

$$x^{(n+1)}(t) = -\int_a^b G^{(n)}(t,\xi) f[\xi, x^{(n)}(\xi), \dot{x}^{(n)}(\xi)]\, d\xi \tag{3.53}$$

instead of (3.44). The Lipschitz-condition for f must now read

$$|f - f^{(n)}| < \alpha_1 |x - x^{(n)}| + \alpha_2 |\dot{x} - \dot{x}^{(n)}|. \tag{3.54}$$

By means of (3.52), (3.53), and (3.54) inequality

$$\begin{aligned} &|x - x^{(n+1)}| \\ &< (b-a) C_1 (\alpha_1 |x - x^{(n)}| + \alpha_2 |\dot{x} - \dot{x}^{(n)}|) + (b-a) |f^{(n)}|_{\max} r^{(n)} \end{aligned} \tag{3.55}$$

is obtained. Since $G(t, \xi)$ is differentiable, $\partial G/\partial t$ exists. Hence, (3.52) yields

$$\dot{x} = -\int_a^b \frac{\partial G(t,\xi)}{\partial t} f[\xi, x(\xi), \dot{x}(\xi)]\, d\xi, \tag{3.56}$$

which is an integral equation for $\dot{x}$. Corresponding to the preceding, the inequality

$$\begin{aligned} &|\dot{x} - \dot{x}^{(n+1)}| \\ &< (b-a) C_2 (\alpha_1 |x - x^{(n)}| + \alpha_2 |\dot{x} - \dot{x}^{(n)}|) + (b-a) |f^{(n)}|_{\max} \bar{r}^{(n)}, \end{aligned} \tag{3.57}$$

where

$$\bar{r}^{(n)} = \left|\frac{\partial G}{\partial t} - \frac{\partial G^{(n)}}{\partial t}\right|_{\max}$$

For $m \geqq 4$, series (3.50) converges for $k = 1$. Therefore, $\bar{r}^{(n)} \to 0$ for $n \to \infty$.

Introducing the norm

$$\|x-x^{(n)}\| = \alpha_1 \,|x-x^{(n)}| + \alpha_2 \,|\dot{x}-\dot{x}^{(n)}|, \tag{3.58}$$

$$\begin{aligned} |x-x^{(n+1)}| &< (b-a)\,C_1 \,\|x-x^{(n)}\| + (b-a)\,|f^{(n)}|_{\max} r^{(n)}, \\ |\dot{x}-\dot{x}^{(n+1)}| &< (b-a)\,C_2 \,\|x-x^{(n)}\| + (b-a)\,|f^{(n)}|_{\max} \bar{r}^{(n)} \end{aligned} \tag{3.59}$$

follows from (3.55), (3.57), finally leading to

$$\begin{aligned} \|x-x^{(n+1)}\| &< (b-a)(\alpha_1 C_1 + \alpha_2 C_2)\,\|x-x^{(n)}\| \\ &\quad + (b-a)\,|f^{(n)}|_{\max}(\alpha_1 r^{(n)} + \alpha_2 \bar{r}^{(n)}). \end{aligned} \tag{3.60}$$

Inequality (3.60) yields after additional $k-1$ steps of iteration

$$\begin{aligned} \|x-x^{(n+k)}\| &< [(b-a)(\alpha_1 C_1 + \alpha_2 C_2)]^k \,\|x-x^{(n)}\| \\ &\quad + \frac{1-[(b-a)(\alpha_1 C_1 + \alpha_2 C_2]^k}{1-(b-a)(\alpha_1 C_1 + \alpha_2 C_2)}\,(b-a)M(\alpha_1 r^{(n)} + \alpha_2 \bar{r}^{(n)}), \end{aligned}$$

which proves

$$\|x-x^{(n+k)}\| \to 0 \text{ for } n, k \to \infty,$$

since $r^{(n)} \to 0$, $\bar{r}^{(n)} \to 0$ for $n \to \infty$, and $(b-a)(\alpha_1 C_1 + \alpha_2 C_2) < 1$ is supposed to hold true.

The conclusion is that the convergence of the iterative process and therefore the convergence of the method of Galerkin is ensured *in the sense of the norm* and under the condition

$$\alpha_1 C_1 + \alpha_2 C_2 < \frac{1}{b-a} \,.$$

However, from the convergence according to the norm, i.e.,

$$\|x-x^{(l)}\| \to 0 \text{ for } l \to \infty,$$

follows by virtue of (3.58)

$$\alpha_1 \,|x-x^{(l)}| + \alpha_2 \,|\dot{x}-\dot{x}^{(l)}| \to 0 \text{ for } l \to \infty.$$

Due to the fact that α_1 and α_2 are constants, the last statement implies that there is convergence in the ordinary sense for x as well as for $\dot{x}$.

Everything else, like convergence toward the fixed point of (3.52) etc., can be shown as before and has therefore not to be discussed in detail anymore.

4

Connection with other methods

4.1 Method of moments

Let

$$Sy + \lambda Ny = 0, \tag{4.1}$$
$$(Gy)_R = 0$$

be the problem under consideration, where S is a selfadjoint positive definite operator, N a nonselfadjoint operator and λ an eigenvalue. The second line in (4.1) represents the boundary conditions of the problem.

As already stated in Section 1.5, an auxiliary problem (1.113), (1.114) is chosen which, with $S \equiv H$, reads

$$\begin{aligned} S\phi_i - \lambda_i \phi_i &= 0, \\ (G\phi_i)_R &= 0. \end{aligned} \tag{4.2}$$

It shall serve to generate a system of functions ϕ_i which is complete with respect to the solutions of (4.1). By virtue of the properties which the operator S has by definition, the operator S has an inverse which reads

$$S^{-1} u(x) = \int_0^l G(x, \xi)\, u(\xi)\, d\xi. \tag{4.3}$$

In (4.3), $0 \leqq x \leqq l$ is the domain of integration of the system, and $G(x, \xi)$ is a symmetric, positive definite Green function, which admits of the bilinear expansion

$$G(x, \xi) = \sum_i \frac{\phi_i(x)\phi_i(\xi)}{\lambda_i}. \tag{4.4}$$

The first line of (4.1) is multiplied by S^{-1} yielding

$$S^{-1}Sy = -\lambda S^{-1}Ny$$

or

$$y(x) = -\lambda \int_0^l G(x,\xi)\,Ny(\xi)\,d\xi. \tag{4.5}$$

The method of moments [27] consists in using for y the series

$$y = \sum_i a_i \phi_i \tag{4.6}$$

in (4.5). This is admissible since the absolute and uniform convergence of (4.6) is known [25]. By means of (4.6), equation (4.5) changes into

$$\sum_i a_i \phi_i(x) = -\lambda \int_0^l G(x,\xi)\,N\left(\sum_i a_i \phi_i(\xi)\right) d\xi. \tag{4.7}$$

Since the operator N is linear, (4.7) can be rewritten to yield

$$\sum_i a_i \phi_i(x) = \sum_i a_i \left[-\lambda \int_0^l G(x,\xi)\,N\phi_i(\xi)\,d\xi\right]. \tag{4.8}$$

Multiplying both sides in (4.8) by $\phi_k(x)$ and integrating over x yields

$$\sum_i a_i \int_0^l \phi_i(x)\phi_k(x)\,dx = \sum_i a_i \left[-\lambda \int_0^l\int_0^l G(x,\xi)N\phi_i(\xi)\phi_k(x)\,d\xi\,dx\right]. \tag{4.9}$$

Due to (4.4) this is equivalent to

$$\sum_i a_i \int_0^l \phi_i(x)\phi_k(x)\,dx$$

$$= \sum_i a_i \left[-\lambda \int_0^l\int_0^l \sum_i \frac{\phi_j(x)\phi_j(\xi)}{\lambda_j}\, N\phi_i(\xi)\phi_k(x)\,d\xi\,dx\right].$$

Integrating on the right hand side over x and using the orthonormality of the coordinate functions ϕ_i yields

$$\sum_i a_i \int_0^l \phi_i(x)\phi_k(x)\,dx = \sum_i a_i \left[-\frac{\lambda}{\lambda_k} \int_0^l \phi_k(\xi) N\phi_i(\xi)\,d\xi \right], \tag{4.11}$$

$$\sum_i a_i \left[\delta_{ik} + \frac{\lambda}{\lambda_k} \int_0^l \phi_k(\xi) N\phi_i(\xi)\,d\xi \right] = 0, \tag{4.12}$$

respectively. These are the equations rendered by the methods of moments to be used to evaluate the eigenvalue λ and the coefficients a_i in the expansion (4.6).

Instead of proceeding further in that direction, it shall rather be shown that these equations are equivalent to those of the method of Galerkin. For this purpose, (4.11) shall be used. Both sides of this relationship are multiplied by λ_k. The result is

$$\sum_i a_i \int_0^l \phi_i(x)\,\lambda_k \phi_k(x)\,dx = \sum_i a_i \left[-\lambda \int_0^l \phi_k(\xi) N\phi_i(\xi)\,d\xi \right]. \tag{4.13}$$

Taking the first line in (4.2) into account, $\lambda_k\, \phi_k(x) = S\phi_k(x)$ is obtained by virtue of which (4.13) changes into

$$\sum_i a_i \int_0^l \phi_i(x)\, S\phi_k(x)\,dx = \sum_i a_i \left[-\lambda \int_0^l \phi_k(\xi) N\phi_i(\xi)\,d\xi \right]. \tag{4.14}$$

Now, the variable of integration ξ may be renamed into x on the right hand side of (4.14), and on the left hand side, ϕ_i may be interchanged with ϕ_k and vice versa. This leads to

$$\sum_i a_i \int_0^l \phi_k(x) S\phi_i(x)\,dx = \sum_i a_i \left[-\lambda \int_0^l \phi_k(x) N\phi_i(x)\,dx \right]. \tag{4.15}$$

A simple transformation of (4.15) yields finally

$$\sum_i a_i \left[\int_0^l \{ S\phi_i(x) + \lambda N\phi_i(x) \} \phi_k(x)\,dx \right] = 0. \tag{4.16}$$

These are the equations of Galerkin corresponding to system (4.1). Consequently, the method of moments and the method of Galerkin prove to be equivalent, and it is apparent that these two methods differ only in their formal appearance. But then, also the convergence problem of the method of moments is settled. The convergence of the method of Galerkin has been established in Section 2.1 and 2.2 using its equivalence to the method of Ritz. But due to the just uncovered equivalence between the method of moments and the method of Galerkin, the method of moments will converge whenever the method of Galerkin converges.

4.2 Method of harmonic balance

Consider

$$\ddot{x} + \omega_0^2 x = f(x), \tag{4.17}$$

$$x(0) = x\left(\frac{2\pi}{\omega}\right), \quad \dot{x}(0) = \dot{x}\left(\frac{2\pi}{\omega}\right), \tag{4.18}$$

where

$$x = x(t), \dot{x} = \frac{dx}{dt}, \omega_0 = \text{const.}, \quad \omega = \text{const.},$$

and where $f(x)$ is a nonlinear function which is continuous and bounded in the whole interval $0 \leq t \leq 2\pi/\omega$. Using the complete system of functions [31]

$$x_k^{(1)} = \sqrt{\frac{\omega}{\pi}} \sin k\omega t, \quad x_k^{(2)} = \sqrt{\frac{\omega}{\pi}} \cos k\omega t, \tag{4.19}$$

which satisfies the boundary conditions (4.18) and is orthonormal, i.e.,

$$\int_0^{2\pi/\omega} x_i^{(r)} x_i^{(s)}\, dt = \delta_{ij}\delta_{rs}, \tag{4.20}$$

one can represent x by means of the series

$$x(t) = \sum_{k=1}^{\infty} a_k^{(1)} x_k^{(1)} + a_k^{(2)} x_k^{(2)}. \tag{4.21}$$

Using this series in (4.17) yields in the usual way the equations of Galerkin

$$\int_0^{2\pi/\omega} \{\sum_k (\omega_0^2 - k^2\omega^2)(a_k^{(1)} x_k^{(1)} + a_k^{(2)} x_k^{(2)})$$

$$- f[\sum_k (a_k^{(1)} x_k^{(1)} + a_k^{(2)} x_k^{(2)})]\} x_j^{(1)} dt = 0, \qquad 4.22)$$

$$\int_0^{2\pi/\omega} \{\sum_k (\omega_0^2 - k^2\omega^2)(a_k^{(1)} x_k^{(1)} + a_k^{(2)} x_k^{(2)})$$

$$- f[\sum_k (a_k^{(1)} x_k^{(1)} + a_k^{(2)} x_k^{(2)})]\} x_j^{(2)} dt = 0.$$

These can be rewritten as

$$(\omega_0^2 - j^2\omega^2)a_j^{(1)} = \int_0^{2\pi/\omega} f[x(\xi)]x_j^{(1)}(\xi)\,d\xi,$$

$$(\omega_0^2 - j^2\omega^2)a_j^{(2)} = \int_0^{2\pi/\omega} f[x(\xi)]x_j^{(2)}(\xi)\,d\xi \tag{4.23}$$

applying (4.20). On the right hand sides of (4.23), x must be replaced by (4.21).
Restricting oneself in (4.23) to the first term in series (4.21) yields

$$a_1^{(1)} = \frac{1}{\omega_0^2-\omega^2}\sqrt{\frac{\omega}{\pi}}\int_0^{2\pi/\omega} f\left[a_1^{(1)}\sqrt{\frac{\omega}{\pi}}\sin\omega\xi + a_1^{(2)}\sqrt{\frac{\omega}{\pi}}\cos\omega\xi\right]\sin\omega\xi\,d\xi,$$

$$a_1^{(2)} = \frac{1}{\omega_0^2-\omega^2}\sqrt{\frac{\omega}{\pi}}\int_0^{2\pi/\omega} f\left[a_1^{(1)}\sqrt{\frac{\omega}{\pi}}\sin\omega\xi + a_1^{(2)}\sqrt{\frac{\omega}{\pi}}\cos\omega\xi\right]\cos\omega\xi\,d\xi. \tag{4.24}$$

Using the new notations

$$a_1^{(1)*} = \sqrt{\frac{\omega}{\pi}}\,a_1^{(1)}, \quad a_1^{(2)*} = \sqrt{\frac{\omega}{\pi}}\,a_1^{(2)}, \tag{4.25}$$

equations (4.24) change into

$$a_1^{(1)*} = \frac{1}{\omega_0^2-\omega^2}\,\frac{\omega}{\pi}\int_0^{2\pi/\omega} f[a_1^{(1)*}\sin\omega\xi + a_1^{(2)*}\cos\omega\xi]\sin\omega\xi\,d\xi,$$

$$a_1^{(2)*} = \frac{1}{\omega_0^2-\omega^2}\,\frac{\omega}{\pi}\int_0^{2\pi/\omega} f[a_1^{(1)*}\sin\omega\xi + a_1^{(2)*}\cos\omega\xi]\cos\omega\xi\,d\xi. \tag{4.26}$$

This is the set of equations for the *method of harmonic balance* as it has been described for example in [32] and [33]. Using for instance $f(x) = x^3$, equations (4.26) yield after a simple calculation

$$(a_1^{(1)*})^2 + (a_1^{(2)*})^2 = \frac{4}{3}(\omega_0^2 - \omega^2).$$

Obviously the choice of one of the coefficients remains arbitrary. Usually, $a_1^{(2)*} = 0$ is chosen. K. Nixdorff has discussed this case in detail [34]. Then, one arrives at the well known result

$$(a_1^{(1)*})^2 = \frac{4}{3}(\omega_0^2 - \omega^2),$$

which gives the relationship between the amplitude $a_1^{(1)*}$ and the frequence ω of the oscillation $x = a_1^{(1)*} \sin \omega t$, which follows from (4.17), (4.18) for $f(x) = x^3$.

As follows from these deliberations, the method of harmonic balance is no more than the first step of approximation of the method of Galerkin using the specific system of functions (4.19). The method of harmonic balance is well founded if it can be shown that the method of Galerkin is convergent when applied to the nonlinear problem (4.17), (4.18), using the system of coordinate functions (4.19). This proof shall be given in the following.

Inserting (4.23) in (4.21) yields

$$x(t) = \sum_{k=1}^{\infty} \frac{1}{\omega_0^2 - k^2\omega^2} \left\{ \left[\int_0^{2\pi/\omega} f[x(\xi)] x_k^{(1)}(\xi)\, d\xi \right] x_k^{(1)}(t) + \left[\int_0^{2\pi/\omega} f[x(\xi)] x_k^{(2)}(\xi)\, d\xi \right] x_k^{(2)}(t) \right\}. \tag{4.27}$$

The Fourier series (4.21) underlying (4.27) is uniformly convergent. Therefore, summation and integration may be interchanged in (4.27). Thus, (4.27) changes after a simple transformation into

$$x(t) = -\int_0^{2\pi/\omega} \sum_{k=1}^{\infty} \frac{x_k^{(1)}(\xi) x_k^{(1)}(t) + x_k^{(2)}(\xi) x_k^{(2)}(t)}{k^2\omega^2 - \omega_0^2} f[x(\xi)]\, d\xi,$$

$$x(t) = -\frac{\omega}{\pi} \int_0^{2\pi/\omega} \sum_{k=1}^{\infty} \frac{\cos k\omega(t-\xi)}{k^2\omega^2 - \omega_0^2} f[x(\xi)]\, d\xi, \tag{4.28}$$

respectively. That is a nonlinear integral equation valid in the regular case $\omega_0 \neq k\omega$. For the nonregular case $\omega_0 = k\omega$, a corresponding integral equation has been derived by K. Nixdorff [35].

It is obvious that the successive approximation of the solution of the nonlinear system of equations (4.23), in which the method of Galerkin consists, is equivalent to the successive approximation of the solution x of the nonlinear integral equation (4.28). The justification of the method of Galerkin and therefore of the method of harmonic balance is thus reduced to the proof of convergence of the successive approximation applied to integral equation (4.28).

This convergence exists under certain assumptions as will be shown.

The integral equation (4.28) can be written as $x = Tx$ using the integral operator

$$Tx(t) = -\frac{\omega}{\pi} \int_0^{2\pi/\omega} K(t, \xi) f[x(\xi)]\, d\xi, \tag{4.29}$$

where $K(t, \xi)$ is the kernel

$$K(t, \xi) = \sum_{k=1}^{\infty} \frac{\cos k\omega (t-\omega)}{k^2 \omega^2 - \omega_0^2}. \tag{4.30}$$

It is continuous and symmetric. Also it is bounded, since

$$|K(t, \xi)| < \frac{1}{\omega^2} \sum_{k=1}^{\infty} \frac{1}{\left| k^2 - \left(\frac{\omega_0}{\omega}\right)^2 \right|}. \tag{4.31}$$

On the right hand side there is a series with positive terms. These terms approach asymptotically the terms of the convergent series

$$\sum_{k=1}^{\infty} \frac{1}{k^2} = \frac{\pi^2}{6}.$$

Then, also series (4.31) converges. Therefore,

$$|K(t, \xi)| < C_1, \tag{4.32}$$

where C_1 is an appropriately chosen constant.

Let Ω be the set of all continuous, bounded and square integrable functions. It is assumed that $f[x(\xi)]$ is bounded. Then, the fixed point

$$x_0 = Tx_0 \tag{4.33}$$

also belongs to Ω, i.e., $x_0 \in \Omega$. Now, let the subset $\Omega_0 \subset \Omega$ be given by

$$x = \{x \in \Omega_0 : |x - x_0| < \delta\}, \quad \delta \geq 0, \quad x_0 \in \Omega_0. \tag{4.34}$$

This is the set of all continuous, bounded and square integrable functions in the neighbourhood of x_0.

The assumed boundedness of $f(x)$ be ensured by the condition

$$f(x) = \epsilon \phi(x), \quad |f|_{max} \leqq \epsilon C_2 \tag{4.35}$$

for $x(\xi) \in \Omega_0$ and $0 \leq \xi \leq 2\pi/\omega$, where ϵ is positive and sufficiently small. Moreover, f be Lipschitz-bounded on Ω_0 so that

$$|f(x)-f(\bar{x})| \leqq \epsilon\alpha\,|x-\bar{x}|\ \text{for}\ x,\bar{x}\in\Omega_0 \tag{4.36}$$

holds true.

It will now be shown that the mapping produced by operator T does not lead outside of Ω_0 if the above made assumptions on K and f are valid. That means, under these assumptions

$$u = Tx\in\Omega_0\ \text{if}\ x\in\Omega_0. \tag{4.37}$$

The correctness of this assertion is proved as follows:

$$|u(t_1)-u(t_2)| \leq 2\,|K(t_1,\xi)-K(t_2,\xi)|\cdot|f|_{\max}.$$

However, K is continuous by assumption, and f is bounded. Therefore,

$$|K(t_1,\xi)-K(t_2,\xi)| \leqq \eta,$$

$$|f|_{\max} \leqq \epsilon C_2.$$

Hence,

$$|u(t_1)-u(t_2) \leqq 2\epsilon\eta C_2 < \tilde{\epsilon},$$

which proves the continuity of u by virtue of the arbitrary smallness of $\tilde{\epsilon}$. Moreover, this continuity is a *uniform* one since $\tilde{\epsilon}$ does not depend on x.

In addition,

$$|u| \leq 2\,|K|\cdot|f|_{\max},$$

which changes into

$$|u| \leqq 2\epsilon C_1 C_2$$

by virtue of (4.32), (4.35), thus providing the *uniform* boundedness of u.

Furthermore,

$$\int_0^{2\pi/\omega} u^2\,dt = \int_0^{2\pi/\omega}\int_0^{2\pi/\omega} \frac{\omega^2}{\pi^2}\,K^2(t,\xi)f^2[x(\xi)]\,d\xi\,dt,$$

$$\left|\int_0^{2\pi/\omega} u^2\,dt\right| < 4\epsilon^2 C_1^2 C_2^2.$$

Hence, u is square integrable.

Finally,

$$|u-x_0| = \left| \frac{\omega}{\pi} \int_0^{2\pi/\omega} K(t,\xi) f[x_0(\xi)]\,d\xi - \frac{\omega}{\pi} \int_0^{2\pi/\omega} K(t,\xi) f[x(\xi)]\,d\xi \right|,$$

which changes into

$$|u-x_0| \leq 2C_1 \epsilon\alpha |x_0-x|$$

using (4.32), (4.36). Since $x \in \Omega_0$ by assumption, (4.34) applies and

$$|u-x_0| \leq 2\epsilon C_1 \alpha\delta \tag{4.38}$$

results. Requiring that

$$\epsilon < \frac{1}{2C_1\alpha}, \tag{4.39}$$

which is possible due to the arbitrariness of ϵ, (4.38) changes into

$$|u-x_0| < \delta.$$

This inequality proves that indeed $u \in \Omega_0$.

Since the operator T maps the set Ω_0 uniformly boundedly and uniformly continuously into itself, this mapping is compact according to the theorem of Arzela [36]. Then, T possesses at least one fixed point according to the theorem of Schauder [37]. Therefore, the assertion that $x_0 \in \Omega_0 \subset \Omega$, made previously, is justified. That means, the integral equation (4.28), and therefore the nonlinear system of equations (4.23), has at least one solution, and the representation of the unknown function by means of the series (4.21) is meaningful.

Now, let it be shown that the solution of (4.28) can be obtained by means of a process of successive approximations. For this purpose, the fixed point theorem of Banach will be applied.

The approximative equation

$$x^{(n)}(t) = -\frac{\omega}{\pi} \int_0^{2\pi/\omega} \sum_{k=1}^{n} \frac{\cos k\omega(t-\xi)}{k^2\omega^2-\omega_0^2} f\left[\sum_{k=1}^{n} a_{k,n}^{(1)} x_k^{(1)}(\xi) + a_{k,n}^{(2)} x_k^{(2)}(\xi) \right] d\xi \tag{4.40}$$

corresponds to integral equation (4.28). It is obtained in the same manner as the exact equation (4.28) follows from the infinite system (4.23) by means of

(4.21). Here, the finite, reduced equations

$$a_{j,n}^{(1)}(\omega_0^2-j^2\omega^2)=\int_0^{2\pi/\omega} f\left[\sum_{k=1}^{n} a_{k,n}^{(1)}x_k^{(1)}(\xi)+a_{k,n}^{(2)}x_k^{(2)}(\xi)\right]x_j^{(1)}(\xi)\,d\xi,$$
$$a_{j,n}^{(2)}(\omega_0^2-j^2\omega^2)=\int_0^{2\pi/\omega} f\left[\sum_{k=1}^{n} a_{k,n}^{(1)}x_k^{(1)}(\xi)+a_{k,n}^{(2)}x_k^{(2)}(\xi)\right]x_j^{(2)}(\xi)\,d\xi, \tag{4.41}$$

which correspond to (4.23), and the finite series development

$$x^{(n)}(t)=\sum_{k=1}^{n}\left[a_{k,n}^{(1)}x_k^{(1)}(t)+a_{k,n}^{(2)}x_k^{(2)}(t)\right] \tag{4.42}$$

are being used. The convergence of the method of Galerkin for this class of problems and, hence, the convergence of the method of harmonic balance is proved, if it is possible to show that the iterative process

$$x^{(n+1)}(t)=-\frac{\omega}{\pi}\int_0^{2\pi/\omega}\sum_{k=1}^{n}\frac{\cos k\omega(t-\xi)}{k^2\omega^2-\omega_0^2}\,f\left[\sum_{k=1}^{n} a_{k,n}^{(1)}x_k^{(1)}(\xi)+a_{k,n}^{(2)}x_k^{(2)}(\xi)\right]d\xi \tag{4.43}$$

leads to a solution x of (4.28) for $n\to\infty$. As an abbreviation set

$$K^{(n)}(t,\xi)=\sum_{k=1}^{n}\frac{\cos k\omega(t-\xi)}{k^2\omega^2-\omega_0^2},$$
$$f^{(n)}(\xi)=f\left[\sum_{k=1}^{n} a_{k,n}^{(1)}x_k^{(1)}(\xi)+a_{k,n}^{(2)}x_k^{(2)}(\xi)\right]. \tag{4.44}$$

Then, (4.28) and (4.43) yield the estimate

$$|x-x^{(n+1)}|\leqq\frac{\omega}{\pi}\int_0^{2\pi/\omega}|Kf-Kf^{(n)}+Kf^{(n)}-K^{(n)}f^{(n)}|\,d\xi.$$

This inequality changes into

$$|x-x^{(n+1)}|<2\epsilon aC_1\,|x-x^{(n)}|+2\,|K-K^{(n)}|\cdot|f^{(n)}|, \tag{4.45}$$

since K is bounded, see (4.32), and f is Lipschitz-bounded, thus satisfying

$$|f-f^{(n)}|_{\max}\leqq\epsilon\alpha\,|x-x^{(n)}|_{\max}.$$

Here, and in the following, all the absolute values used shall be by convention the largest possible in the interval $0 \leq \xi \leq 2\pi/\omega$. Because of the convergence of (4.30), $K^{(n)} \to K$ for $n \to \infty$, and

$$|K - K^{(n)}| = r^{(n)} \to 0 \text{ for } n \to \infty.$$

From (4.45),

$$|x - x^{(n+k)}| < (2\epsilon\alpha C_1)^k |x - x^{(n)}| + \frac{1-(2\epsilon\alpha C_1)^k}{1-2\epsilon\alpha C_1} 2Mr^{(n)} \tag{4.46}$$

is obtained after $k-1$ additional steps of iteration. M is the maximum of all $|f^{(p)}|_{\max}, p = n, n+1, \ldots, n+k-1$. Under the assumption (4.39),

$$(2\epsilon\alpha C_1)^k \to 0 \text{ for } k \to \infty.$$

Furthermore, $r^{(n)} \to 0$ for $n \to \infty$. Consequently, (4.46) leads to

$$|x - x^{(n+k)}| \to 0 \text{ for } n, k \to \infty.$$

This proves the convergence of the iterative process (4.43). Hence, the convergence of the method of Galerkin and of the method of harmonic balance is proved.

Condition (4.39) is sufficient for the convergence being toward the fixed point x_0 in Ω_0. If the convergence of (4.43) were toward $\tilde{x} \in \Omega_0$ then, because of (4.33),

$$|x_0 - \tilde{x}| < 2\epsilon\alpha C_1 \, |x_0 - \tilde{x}|.$$

This is in contradiction to (4.39) for $x_0 \neq \tilde{x}$. Hence, $x_0 \equiv \tilde{x}$.

The previous results may be summarized as follows:

For the problem.

$$\ddot{x} + \omega_0^2 x = f(x) = \epsilon\phi(x),$$

$$x(0) = x\left(\frac{2\pi}{\omega}\right), \quad \dot{x}(0) = \dot{x}\left(\frac{2\pi}{\omega}\right),$$

a periodic solution having the period ω is sought by means of the method of harmonic balance which is considered to be a first approximation in the sense of the method of Galerkin. The nonlinear function $f(x)$ be continuous, bounded and Lipschitz-bounded for $x \in \Omega_0$. Then, a periodic solution $x_0 \in \Omega_0$ exists, and the iterative process (4.43) obtained as a generalization of the method of harmonic balance converges toward x_0 if

$$\epsilon < \frac{2}{2C_1 \alpha}.$$

Here, C_1 is the maximum value of the series

$$\sum_{k=1}^{\infty} \frac{\cos k\omega(t-\xi)}{k^2\omega^2-\omega_0^2}$$

which the series assumes for $\omega_0 \neq k\omega$, and α is the Lipschitz constant of ϕ on Ω_0.

An extension of this theory for the case that the right hand side of (4.17) consists of $f(x)$ instead of $f(x, \dot{x})$, and for the case that the left hand side of (4.17) consists of the linear differential operator

$$\sum_{i=0}^{m} C_i \frac{d^i x}{dt^i}, \quad C_i = \text{const}$$

instead of $\ddot{x} + \omega_0^2$ can be found in [38].

4.3 Method of Grammel

Consider the boundary value problem

$$Sy - \lambda my = 0, \tag{4.47}$$

$$(Gy)_R = 0, \tag{4.48}$$

where S is a positive definite, selfadjoint differential operator, λ an eigenvalue, m a function which is positive in the interval $a \leqq x \leqq b$, and where the end points $x = a$ and $x = b$ of the x-interval represent the boundary R of the system. The boundary conditions of the problem are given by (4.48).

Let

$$S\phi_i - \lambda\phi_i = 0, \tag{4.49}$$

$$(G\phi_i)_R = 0 \tag{4.48}$$

be the auxiliary problem generating the coordinate functions ϕ_i. For these coordinate functions,

$$\int_a^b \phi_i \phi_j = \delta_{ij}, \quad \int_a^b \phi_j S\phi_i dx = \lambda_i \delta_{ij}. \tag{4.50}$$

Functions ϕ_i may be used to construct the Green function of operator S which satisfies (4.48) and admits of the bilinear expansion

$$G(x, \xi) = \sum_k \frac{\phi_k(x)\,\phi_k(\xi)}{\lambda_k} \tag{4.51}$$

These functions may also be used to expand the solution of (4.47) into the series

$$y = \sum_i a_i \phi_i. \tag{4.52}$$

Following this, the *equations of Galerkin*

$$\int_a^b [S \sum_i a_i \phi_i - \lambda m \sum_i a_i \phi_i]\,\phi_k\, d\xi = 0 \tag{4.53}$$

of problem (4.47), (4.48) may be obtained in the usual way. They can be re-written by using the second equation in (4.50) to yield

$$a_k \lambda_k - \int_a^b m \sum_i a_i \phi_i \phi_k\, d\xi = 0. \tag{4.54}$$

Multiplying both sides of (4.54) by $\phi_k(x)$, dividing by λ_k and then adding over k yields

$$\sum_k a_k \phi_k(x) - \lambda \int_a^b m(\xi) \left(\sum_i a_i \phi_i(\xi)\right) \sum_k \frac{\phi_k(x)\,\phi_k(\xi)}{\lambda_k}\, d\xi = 0.$$

Because of (4.51), this relationship is equivalent to

$$\sum_k a_k \phi_k(x) - \lambda \int_a^b m(\xi)\, G(x, \xi) \left(\sum_i a_i \phi_i(\xi)\right) d\xi = 0. \tag{4.55}$$

This is an equation which can also be obtained by applying the *method of moments* to the integral equation

$$y(x) - \lambda \int_a^b G(x, \xi)\, m(\xi)\, y(\xi)\, d\xi = 0 \tag{4.56}$$

which corresponds to (4.47), (4.48).

Relationship (4.55) may be rewritten to yield

$$\sum_k a_k \left[\phi_k(x) - \lambda \int_a^b m(\xi)\, G(x,\xi)\, \phi_k(\xi)\, d\xi \right] = 0. \tag{4.57}$$

Since the expression in the square bracket must be identical zero, it can be multiplied by $\phi_i(x)$ and integrated over x, and this operation does not change anything. The result is

$$\sum_k a_k \left\{ \int_a^b \left[\phi_k(x) - \lambda \int_a^b m(\xi)\, G(x,\xi)\, \phi_k(\xi)\, d\xi \right] \phi_i(x)\, dx \right\} = 0. \tag{4.58}$$

One is thus arrived at the *equations of Grammel* [41].

If the equations of Galerkin and the equivalent equations of the method of moments (see Section 4.1) are said to be "equations of the first order", then, the equations of Grammel must be considered to be "equations of the second order" since they involve one or more integration as the preceding equations. By this repeated integration, the accuracy in the approximate numerical evaluation of the eigenvalue λ is increased. This fact has already been pointed out by Grammel himself.

The convergence of the method of Grammel can be proved under wider assumptions as Grammel himself had introduced. This will be shown now. For this purpose consider

$$Ly - \lambda m y = 0, \tag{4.59}$$

$$(Gy)_R = 0, \tag{4.48}$$

where everything is as before, except that the linear differential operator is allowed to be nonselfadjoint. In addition, the following assumptions are made:

(i) The operator L has the inverse

$$L^{-1} = \int_a^b G(x,\xi) \ldots d\xi,$$

where $G(x,\xi)$ is continuous, bounded, but not necessarily symmetric.

(ii) The function m be positive and bounded in the interval $a \leqq x \leqq b$.

(iii) A complete and in the general sense orthogonal system of functions ϕ_i is available for which therefore

$$\int_a^b \left[y - \sum_{i=1}^{n} \phi_i \right]^2 dx = \epsilon^2, \quad \epsilon \to 0 \text{ for } n \to \infty, \tag{4.60}$$

and

$$\int_a^b m(x)\,\phi_i(x)\,\phi_j(x)\,dx = \delta_{ij} \tag{4.61}$$

holds true.

An example on how to generate such a system ϕ_i is the following: Choose the auxiliary problem

$$S\phi_i - \lambda_i m\phi_i = 0, \tag{4.62}$$

$$(G\phi_i)_R = 0, \tag{4.48}$$

which possesses the eigenfunctions ϕ_i and the Green function $G^*(x, \xi)$. The integral equation

$$\phi_i(x) = \lambda_i \int_a^b m(\xi)\,G^*(x, \xi)\,\phi_i(\xi)\,d\xi \tag{4.63}$$

is equivalent to (4.62), (4.48). Introducing the functions

$$\psi_i(x) = \sqrt{m(x)}\,\phi_i(x), \tag{4.64}$$

and multiplying both sides of (4.63) by $\sqrt{m(x)}$ yields

$$\psi_i(x) = \lambda_i \int_a^b \sqrt{m(\xi)}\sqrt{m(x)}\,G^*(x, \xi)\,\psi_i(\xi)\,d\xi. \tag{4.65}$$

This is an integral equation having the *symmetric* kernel $K(x, \xi) = \sqrt{m(\xi)}\sqrt{m(x)}\,G^*(x, \xi)$. Therefore, the eigenfunctions ψ_i of this equation are orthonormal satisfying the condition

$$\int_a^b \psi_i \psi_k\,dx = \delta_{ik}. \tag{4.66}$$

Using in (4.66) the relationship (4.64) yields

$$\int_a^b m(x)\,\phi_i(x)\,\phi_k(x)\,dx = \delta_{ik}, \tag{4.67}$$

which is the required orthogonality in the general sense (4.61).

Since Green's function $G(x, \xi)$ of (4.59), (4.48) has been assumed to be continuous and bounded, the "Fourier series"

$$G(x,\xi) = \sum_{i,r} \gamma_{ir}\phi_i(x)\,\phi_r(\xi) \tag{4.68}$$

applies at least formally. From (4.61) and (4.68) follows that

$$\int_a^b G(x,\xi)\,m(x)\,\phi_k(x)\,dx = \sum_r \gamma_{kr}\phi_r(\xi) \tag{4.69}$$

is the k-th "Fourier coefficient" of $G(x, \xi)$ in the expansion (4.68) with respect to the $\phi_i(x)$. For (4.69) at least Bessel's inequality

$$\sum_k \left(\sum_r \gamma_{kr}\phi_r(\xi)\right)^2 \leqq \int_a^b G^2(x,\xi)\,m(x)\,m(\xi)\,dx\,d\xi \tag{4.70}$$

holds true. Integrating (4.70) yields

$$\int_a^b \sum_k \left(\sum_r \gamma_{kr}\,\phi_r(\xi)\right)^2 m(\xi)\,d\xi \leqq \int_a^b\int_a^b G^2(x,\xi)\,m(x)\,m(\xi)\,dx\,d\xi$$

from which by virtue of (4.61) and of

$$\int_a^b\int_a^b G^2(x,\xi)\,m(\xi)\,dx\,d\xi \leqq P \tag{4.71}$$

$$\sum_{i,k} \gamma_{ik}^2 \leqq P \tag{4.72}$$

is obtained. These steps of calculation are based on the continuity and boundedness of $G(x, \xi)$. According to (4.72), the double series $\sum_{i,k} \gamma_{ik}^2$ converges. Then, also the simple sum $\sum_k \gamma_{ik}^2$ converges. According to the theorem of Riesz-Fischer [42], there exists the square integrable function

$$u_i(\xi) = \sum_r \gamma_{ir}\phi_r(\xi). \tag{4.73}$$

By means of this function, (4.68) and (4.69) can be rewritten to yield

$$G(x,\xi) = \sum_i u_i(\xi)\,\phi_i(x) \tag{4.74}$$

and

$$u_i(\xi) = \int_a^b G(x,\xi)\, m(x)\, \phi_i(x)\, dx. \tag{4.75}$$

Finally, a statement concerning the coefficients γ_{ii} of (4.68) may be made: since (4.72) applies, the coefficients

$$\gamma_{ii} = \int_a^b \int_a^b G(x,\xi)\, m(x)\, \phi_i(x)\, m(\xi)\, \phi_i(\xi)\, dx\, d\xi \tag{4.76}$$

must have decreasing values for increasing indices i.

Next, it may be shown that it is justified in the sense of the method of Grammel to work with the finite system of equations

$$\sum_{k=1}^{n} \left\{ \int_a^b \left[\phi_k(x) - \bar{\lambda} \int_a^b m(\xi) G(x,\xi)\, \phi_k(\xi)\, d\xi \right] m(x)\, \phi_i(x)\, dx \right\} = 0 \tag{4.77}$$

instead of the infinite system

$$\sum_{k=1}^{\infty} \left\{ \int_a^b \left[\phi_k(x) - \lambda \int_a^b m(\xi) G(x,\xi)\, \phi_k(\xi)\, d\xi \right] m(x)\, \phi_i(x)\, dx \right\} = 0. \tag{4.78}$$

That means, it may be shown that the method of reduction is admissible. For this purpose, (4.68) may be inserted into (4.77), (4.78), respectively. Using (4.67) and carrying out some intermediate calculations, the finite, infinite system of algebraic equations

$$\sum_{k=1}^{n} a_k \left[\delta_{ik} - \bar{\lambda} \gamma_{ik}\right] = 0 \qquad i = 1,2,\ldots,n, \tag{4.79}$$

$$\sum_{k=1}^{\infty} a_k \left[\delta_{ik} - \lambda \gamma_{ik}\right] = 0, \qquad i = 1,2,3,\ldots, \tag{4.80}$$

is obtained, respectively. According to L. W. Kantorowitch and G. P. Akilow [43], the method of reduction applies to (4.80), if the coefficients in (4.80) satisfy the condition $\sum_{i,k} \gamma_{ik}^2 < \infty$. Then, the infinite system (4.80) may be replaced by the finite system (4.79). In (4.72) it has been shown that the condition for the coefficients is satisfied. Hence, the use of the finite system of equations (4.77), and thus, the method of Grammel is justified.

It remains to be proven that the approximate value $\bar{\lambda}$, obtained by means of (4.77), tends toward the exact value λ. Because of (4.56),

$$\int_a^b \left[y(x) - \lambda \int_a^b G(x,\xi) y(\xi) m(\xi) d\xi \right] m(x) \phi_i(x) dx = 0, \tag{4.81}$$

$i = 1,2,3, \ldots,$

is an exact statement. In its place,

$$\int_a^b \left[y^{(n)}(x) - \bar{\lambda} \int_a^b G(x,\xi) y^{(n)}(\xi) m(\xi) d\xi \right] m(x) \phi_i(x) dx = 0, \tag{4.82}$$

$i = 1,2, \ldots, n,$

is being used as an approximation in the context of the method of Grammel. In (4.82),

$$y^{(n)}(x) = \sum_{i=1}^{n} a_i \phi_i(x). \tag{4.83}$$

Subtracting (4.81) from (4.82) and adding

$$\int_a^b \left[\lambda \int_a^b G(x,\xi) y^{(n)}(\xi) m(\xi) d\xi - \lambda \int_a^b G(x,\xi) y^{(n)}(\xi) m(\xi) d\xi \right] m(x) \phi_i(x) dx,$$

which does not change anything, one arrives at

$$\begin{aligned} &\int_a^b \{[y^{(n)}(x) - y(x)] - [\bar{\lambda} - \lambda] \int_a^b G(x,\xi) y^{(n)}(\xi) d\xi \\ &\quad - \lambda \int_a^b G(x,\xi) m(\xi) [y^{(n)}(\xi) - y(\xi)] d\xi\} m(x) \phi_i(x) dx = 0. \end{aligned} \tag{4.84}$$

This equation can be changed into

$$\begin{aligned} &[\bar{\lambda} - \lambda] \int_a^b \int_a^b G(x,\xi) m(x) \phi_i(x) m(\xi) y^{(n)}(\xi) dx\, d\xi \\ &= \int_a^b [y^{(n)}(x) - y(x)] m(x) \phi_i(x) dx \end{aligned} \tag{4.85}$$

$$-\lambda \int_a^b \int_a^b G(x,\xi) m(x) \phi_i(x) m(\xi) [y^{(n)}(\xi) - y(\xi)] \, dx \, d\xi.$$

Using (4.75) in (4.85) yields

$$[\bar{\lambda} - \lambda] \int_a^b u_i(\xi) y^{(n)}(\xi) m(\xi) \, d\xi$$

$$= \int_a^b [y^{(n)}(x) - y(x)] m(x) \phi_i(x) \, dx - \lambda \int_a^b u_i(\xi) m(\xi) [y^{(n)}(\xi) - y(\xi)] \, d\xi. \tag{4.86}$$

Renaming the variables of integration on the right hand side of (4.86) and using t instead, allows one to rewrite (4.86) in the more concise form

$$[\lambda^{(n)} - \lambda] \int_a^b u_i(\xi) y^{(n)}(\xi) m(\xi) \, d\xi$$

$$= \int_a^b [y^{(n)}(t) - y(t)] [\phi_i(t) - \lambda u_i(t)] m(t) \, dt. \tag{4.87}$$

The inequality of Schwarz yields

$$\left[\int_a^b u_i(\xi) y^{(n)}(\xi) m(\xi) \, d\xi \right]^2$$

$$\leqq \left[\int_a^b u_i^2(\xi) m(\xi) \, d\xi \right] \cdot \left[\int_a^b [y^{(n)}(\xi)]^2 m(\xi) \, d\xi \right]. \tag{4.88}$$

Since m is positive and $u_i(\xi)$ is square integrable,

$$\int_a^b u_i^2(\xi) m(\xi) \, d\xi = U, \tag{4.89}$$

where U is an appropriate, finite, positive number. Also, one finds easily that

$$\int_a^b u_i^2(\xi) m(\xi) \, d\xi = \sum_k \gamma_{ik}^2,$$

so that $U = \sum_k \gamma_{ik}^2$. Because of (4.72), $U < P$. Moreover, because of (4.61) and (4.83),

$$\int_a^b [y^{(n)}(\xi)]^2\, m(\xi)\, d\xi = \sum_{i=1}^n a_i^2 = N, \tag{4.90}$$

where N is also an appropriate, finite, positive number. On the basis of (4.89), (4.90), equation (4.88) yields

$$\left[\int_a^b u_i(\xi) y^{(n)}(\xi) m(\xi)\, d\xi\right]^2 = K < PN, \tag{4.91}$$

where K is a third appropriate, finite, positive number.

Now, both sides in (4.87) may be squared. Relationship (4.91) may be applied to the left hand side, and to the right hand side the inequality of Schwarz may be applied. The result is

$$[\lambda^{(n)} - \lambda]^2 K \leqq \int_a^b m(t)\, [y^{(n)}(t) - y(t)]^2\, dt \int_a^b m(t)\, [\phi_i(t) - \lambda u_i(t)]^2\, dt. \tag{4.92}$$

If M is the mean value of $m(t)$, the first factor on the right hand side of (4.92) can be expressed as

$$\int_a^b m(t)\, [y^{(n)}(t) - y(t)]^2\, dt = M \int_a^b [y^{(n)}(t) - y(t)]^2\, dt. \tag{4.93}$$

The second factor can be rewritten using (4.61) and (4.73) to yield

$$\int_a^b m(t)\, [\phi_i(t) - \lambda u_i(t)]^2\, dt = \int_a^b m\phi_i^2\, dt - 2\lambda \int_a^b m\phi_i u_i dt + \lambda^2 \int_a^b m u_i^2\, dt$$

$$= 1 - 2\lambda\gamma_{ii} + \lambda^2 \sum_k \gamma_{ik}^2. \tag{4.94}$$

It is already known that $\sum_k \gamma_{ik}^2 = U$, where U is a finite, positive number. Moreover, due to the convergence stated by (4.72), the coefficients γ_{ii} given by (4.76) must have decreasing values for increasing i, as already mentioned previously. Hence, one can conclude that

$$\int_a^b m(t)[\phi_i(t) - \lambda u_i(t)]^2 \, dt = 1 - 2\lambda\gamma_{ii} + \lambda^2 \sum_k \gamma_{ik}^2 < \Gamma, \tag{4.95}$$

where Γ is a finite, positive number, since

$$1 - 2\lambda\gamma_{ii} + \lambda^2 \sum_k \gamma_{ik}^2 > 1 - 2\lambda\gamma_{ii} + \lambda^2\gamma_{ii}^2 = (1 - \lambda\gamma_{ii})^2 > 0.$$

Using (4.93) and (4.95), inequality (4.92) changes into

$$[\lambda^{(n)} - \lambda]^2 K \leqq M\Gamma \int_a^b [y^{(n)}(t) - y(t)]^2 \, dt. \tag{4.96}$$

Because of (4.83) and the completeness of the system ϕ_i, condition (4.60) holds true. Hence, (4.96) yields

$$|\lambda^{(n)} - \lambda| \leqq \left(\frac{M\Gamma}{K}\right)^{1/2} \epsilon. \tag{4.97}$$

Since $\epsilon \to 0$ for $n \to \infty$, finally,

$$|\lambda^{(n)} - \lambda| \to 0 \text{ for } n \to \infty \tag{4.98}$$

follows from (4.97). Thus, it is proved that the approximate value $\lambda^{(n)}$ obtained by means of the method of Grammel converges toward the exact eigenvalue λ of problem (4.59), (4.48).

For the practical application of the method, difficulties may arise from the fact that in the equations of Grammel (4.82), the Green function $G(x, \xi)$ of problem (4.59), (4.48) must be known. Due to possibly complicated boundary conditions (4.48), the knowledge of $G(x, \xi)$ may sometimes be ruled out. Then, the method would not be practicable. Therefore, it may be shown in the following, that there is a way out of these difficulties which is in close connection with the extended equations of Galerkin mentioned in Section 3.3.

The posed problem be again (4.59), (4.48). Its Green function may not be known. Therefore, the "substitute problem"

$$Ly - \lambda m y = 0, \tag{4.99}$$

$$(\bar{G}y)_R = 0 \tag{4.100}$$

may be introduced, in which the operator L be of order μ, and selfadjoint, positive definite with respect to the boundary conditions (4.100), which are a properly modified version of conditions (4.48). The modified conditions (4.100)

be specifically chosen in such a way that the Green function $\bar{G}(x, \xi)$ corresponding to (4.99), (4.100) can be expressed explicitely. Then, the integral equation

$$\psi_i(\xi) = \lambda_i \int_a^b \psi_i(x) m(x) \bar{G}(x, \xi) dx \tag{4.101}$$

is equivalent to the substitute problem. In (4.101), the ψ_i are the eigenfunctions corresponding to the eigenvalues λ_i. These eigenvalues are by the way always positive according to the assumptions made with respect to the substitute problem.

Introducing the new functions

$$\rho_i(\xi) = \sqrt{m(\xi)}\,\psi_i(\xi), \tag{4.102}$$

equation (4.101) can be transformed into

$$\rho_i(\xi) = \lambda_i \int_a^b \rho_i(x) \bar{G}(x, \xi) \sqrt{m(\xi)} \sqrt{m(x)}\, dx. \tag{4.103}$$

In (4.103), the kernel

$$K(x, \xi) = \bar{G}(x, \xi) \sqrt{m(\xi)} \cdot \sqrt{m(x)} \tag{4.104}$$

is *symmetric*, but it is also positive definite under the reigning assumptions. Consequently, the $\rho_i(\xi)$ are the eigenfunctions of

$$\rho_i(\xi) = \lambda_i \int_a^b \rho_i(x) K(x, \xi) dx \tag{4.105}$$

and can therefore be considered to be orthonormal, i.e., satisfying

$$\int_a^b \rho_i(\xi) \rho_k(\xi) d\xi = \delta_{ik}. \tag{4.106}$$

Because of (4.102),

$$\int_a^b m(\xi) \psi_i(\xi) \psi_k(\xi) d\xi = \delta_{ik}, \tag{4.107}$$

i.e., the functions $\psi_i(\xi)$ are orthonormal in the general sense.

Because of its symmetry and positive definitness, the kernel $K(x, \xi)$ is bilinearly expandable, i.e., the series

$$K(x, \xi) = \sum_i \frac{\rho_i(x)\rho_i(\xi)}{\lambda_i} \tag{4.108}$$

converges absolutely and uniformly. Because of (4.102), (4.104), also

$$\bar{G}(x, \xi) = \sum_i \frac{\psi_i(x)\psi_i(\xi)}{\lambda_i}. \tag{4.109}$$

Hence, the Green function $\bar{G}(x, \xi)$ is bilinearly expandable too. By means of (4.109), and (4.107) one finds that in addition

$$\int_a^b \bar{G}(x, x)m(x)dx = \sum_i \frac{1}{\lambda_i}. \tag{4.110}$$

Since the integral on the left hand side of (4.110) represents a finite value,

$$\sum_i \frac{1}{\lambda_i} < \infty.$$

Consequently, this infinite series converges.

In the following, a certain example shall be used to illustrate actually quite general facts and statements. Let $\mu = 4$, i.e., the problem concerned

$$Ly - \lambda my = 0, \tag{4.99}$$

$$y(a) = y'(a) = y''(b) = 0, \quad y'''(b) = \gamma\lambda y'(b) \tag{4.111}$$

be of the fourth order. In (4.111), γ is a constant. The substitute problem be

$$Ly - \lambda my = 0, \tag{4.99}$$

$$y(a) = y'(a) = y''(b) = y'''(b) = 0. \tag{4.112}$$

Through this choice, the fourth, disturbing condition in (4.111) has been removed and replaced by a simpler one. Now, let Green's identity

$$\int_a^b [uL(v) - vL(u)]\,dx = \Big[uv''' - u'v'' + u''v' - u'''v\Big]_a^b \tag{4.113}$$

be especially applied to $u = \bar{G}(x, \xi)$ and $v = y(x)$, where *y is supposed to be the solution of the original problem (4.99), (4.111)*. The result is

$$\int_a^b [\lambda m y(x) \bar{G}(x,\xi) - y(x)\delta(x,\xi)]\,dx \tag{4.114}$$

$$= \left[y'''(x)\bar{G}(x,\xi) - y''(x)\frac{\partial \bar{G}(x,\xi)}{\partial x} + y'(x)\frac{\partial^2 \bar{G}(x,\xi)}{\partial x^2} - y(x)\frac{\partial^3 \bar{G}(x,\xi)}{\partial x^3} \right]_a^b .$$

This identity can be rewritten taking into account that $\bar{G}(x,\xi)$ satisfies the boundary conditions (4.112) and y the boundary conditions (4.111). In this way one arrives at

$$y(\xi) = \lambda \int_a^b \bar{G}(x,\xi) m(x) y(x)\,dx - y'''(b)\bar{G}(b,\xi), \tag{4.115}$$

which is an integral equation for the solution $y(x)$ which is to be determined. It is noteworthy that on the right hand side of (4.115) there appears the "weighing term" $-y'''(b)\bar{G}(b,\xi)$ [44]. Using the last condition in (4.111), equation (4.115) can again be rewritten to yield finally

$$y(\xi) = \lambda \int_a^b \bar{G}(x,\xi) m(x) y(x)\,dx - \gamma\lambda y'(b)\bar{G}(b,\xi). \tag{4.116}$$

Considering this example one can easily state by induction that in the general case, the corresponding application of Green's identity will lead to equations of the form

$$y(\xi) = \lambda \int_a^b \bar{G}(x,\xi) m(x) y(x)\,dx \pm \sum_k A_k y^{(\nu_k)}(R)\,\frac{\partial^k \bar{G}(R,\xi)}{\partial x^k} . \tag{4.117}$$

In (4.117), the A_k are certain constants. The $y^{(\nu_k)}(R)$ are derivatives of y at a point of the boundary R. The order ν of these derivatives may change from term to term. This is indicated by means of the superscript (ν_k) with variable k.

In order to keep things simple, we shall deal with the case

$$y(\xi) = \lambda \int_a^b \bar{G}(x,\xi) m(x) y(x)\,dx - A y^{(\nu)}(R)\,\frac{\partial^k \bar{G}(R,\xi)}{\partial x^k}, \tag{4.118}$$

where there is only a single weighing term present.

Rearranging (4.118) into

$$y(\xi)+Ay^{(\nu)}(R)\frac{\partial^k \bar{G}(R,\xi)}{\partial x^k}=\lambda\int_a^b \bar{G}(x,\xi)m(x)y(x)dx, \tag{4.119}$$

it becomes apparent that the expression

$$y(\xi)+Ay^{(\nu)}(R)\frac{\partial^k \bar{G}(R,\xi)}{\partial x^k} \tag{4.120}$$

is expressible in terms of the kernel on the right hand side of (4.119). Hence, it is expandable in terms of the eigenfunctions $\psi_i(x)$ of $\bar{G}(x,\xi)$. That means,

$$y(\xi)+Ay^{(\nu)}(R)\frac{\partial^k \bar{G}(R,\xi)}{\partial x^k}=\sum_k c_k\,\psi_k(\xi), \tag{4.121}$$

where the c_k are the coefficients of the expansion.

However, $Ay^{(\nu)}(R)$ is a fixed number. Therefore, if $\partial^k\bar{G}(R,\xi)/\partial x^k$ is expandable, i.e., if

$$\frac{\partial^k \bar{G}(R,\xi)}{\partial x^k}=\sum_i\frac{\psi_i^{(k)}(R)\,\psi_i(\xi)}{\lambda_i} \tag{4.122}$$

is abolutely and uniformly convergent, then,

$$y(\xi)=\sum_i\left(c_i-Ay^{(\nu)}(R)\,\frac{\psi_i^{(k)}(R)}{\lambda_i}\right)\psi_i(\xi) \tag{4.123}$$

follows from (4.121) and (4.122). That implies that $y(\xi)$ *itself* is expandable with respect to the $\psi_k(\xi)$, and this expansion converges absolutely and uniformly. This fact is in agreement with the results in Section 2.2, where it has been shown that the solution y of a problem can be developed into a series in terms of coordinate functions ϕ_i which do satisfy the geometric boundary conditions only and violate the dynamic ones. But this is just the case for the coordinate functions ψ_i being used here.

By means of the abbreviation

$$c_i-Ay^{(\nu)}(R)\,\frac{\psi_i^{(k)}(R)}{\lambda_i}=a_i, \tag{4.124}$$

one arrives finally at

$$y(\xi)=\sum_i a_i\,\psi_i(\xi). \tag{4.125}$$

The question is, when it might be admissable to use (4.122). This question can be answered as follows: according to L. Collatz [45],

$$\sum_i \frac{[\psi_i^{(k)}(R)]^2}{\lambda_i} \leqq \left(\frac{\partial^{2k} G(x,\xi)}{\partial x^k \partial \xi^k}\right)_{x=\xi=R} \qquad \text{for } k \leqq \frac{\mu}{2} - 1, \tag{4.126}$$

where the series on the left hand side proves to be convergent. Moreover, because of (4.109),

$$\sum_i \frac{\psi_i^2(\xi)}{\lambda_i} = \bar{G}(\xi,\xi). \tag{4.127}$$

Hence, also this series converges absolutely and uniformly. Now, (4.109) may be used to set up series (4.122) formally. By means of the inequality of Schwarz, one finds that

$$\sum_i \left| \frac{\psi_i^{(k)}(R)}{\sqrt{\lambda_i}} \frac{\psi_i(\xi)}{\sqrt{\lambda_i}} \right| \leqq \left[\sum_i \frac{(\psi_i^{(k)}(R))^2}{\lambda_i}\right]^{1/2} \cdot \left[\sum_i \frac{(\psi_i(\xi))^2}{\lambda_i}\right]^{1/2}. \tag{4.128}$$

The conclusion is that the series (4.122) converges indeed absolutely and uniformly because of (4.126), (4.127), *if* $k \leqq \mu/2 - 1$ holds true. This result implies that under the same condition the series expansion (4.125) for y holds true.

The next question is under which circumstances the weighing terms in (4.117) may involve derivatives of $\bar{G}(R,\xi)$ with respect to x which are of an order $k \leq \mu/2 - 1$. Considering (4.114), it becomes apparent that this is the case if in defining the substitute problem *one violates only boundary conditions which are of a sufficient high order*, i.e., of the order $\mu - 1$ and downward, in the extreme case, to the order $\mu/2$. These are then just the dynamic boundary conditions.

Later on, the condition is needed, that the factor $y^{(\nu)}(R)$ in the weighing term be absolutely and uniformly expandable. Therefore, it may be now investigated when that might be the case. For this purpose

$$y^{(\nu)}(\xi) = \sum_k a_k \psi_k^{(\nu)}(\xi) \tag{4.129}$$

may be derived formally from (4.125). Now consider

$$\int_a^b y\, L d\xi = \int_a^b [\sum_k a_k \psi_k(\xi)]\, L \left(\sum_r a_r \psi_r(\xi)\right) d\xi.$$

This relationship changes into

$$\lambda \int_a^b m(\xi) y^2(\xi)\, d\xi = \sum_k a_k^2 \lambda_k \tag{4.130}$$

because of $Ly = \lambda my$ and condition (4.107). Assuming that $y(\xi)$ be square integrable, it is obvious that the series on the right hand side of (4.130) converges. Now, (4.129) can be estimated according to the inequality of Schwarz. The result is

$$\sum_k a_k \psi^{(\nu)}(\xi) \leqq \left(\sum_k a_k^2 \lambda_k\right)^{1/2} \left(\sum_k \frac{(\psi_k^{(\nu)}(\xi))^2}{\lambda_k}\right)^{1/2} \tag{4.131}$$

Because of (4.126), a relationship which is valid for arbitrary $x = \xi$ and not only for $x = \xi = R$, and of (4.130) it follows from (4.131) that the expansion (4.129) converges absolutely and uniformly for $\nu \leqq \mu/2 - 1$, in which case it represents in fact the derivative $y^{(\nu)}$.

Now, it may be summarized under which conditions the here introduced extension of the method of Grammel will be valid:

(i) The solution y to be determined must be square integrable.

(ii) In defining the substitute problem, only derivatives from the order $\mu - 1$ downward to the order $\mu/2$ should be violated when replacing the boundary conditions $(Gy)_R = 0$ of the actual problem (4.59), (4.48) by the boundary conditions $(\bar{G}y)_R = 0$ of the substitute problem.

(iii) Through the violation of these derivatives, only derivatives $y^{(\nu)}$ of an order $\nu \leqq \mu/2 - 1$ and derivatives $\partial^k \bar{G}/\partial x^k$ of an order $k \leqq \mu/2 - 1$ are supposed to get into the weighing terms of (4.117).

For an illustration of these statements, the previous example may be used: Contrary to the last condition in (4.111), $y'''(b) = 0$ has been assumed in (4.112), i.e., the original condition $y'''(b) = \gamma\lambda y'(b)$ has been violated. That is a derivative of the $(\mu - 1)$th $= (4 - 1)$th $=$ third order. As a consequence of this, the weighing term of the integral equation involves the term $\gamma\lambda y'(b)$ which contains a derivative of the $(\mu/2 - 1)$th $= (4/2 - 1)$th $=$ first order, and a derivative of $\bar{G}$ of zero order. Hence, all conditions previously mentioned are satisfied in this case if in addition y turns out to be square integrable. This is of course the case due to the nature of the problem.

In complete correspondence to the original method of Grammel, the system of equations

$$\int_a^b \left[y(\xi) - \lambda \int_a^b \bar{G}(x,\xi) m(x) y(x)\, dx + A y^{(\nu)}(R) \frac{\partial^k \bar{G}(R,\xi)}{\partial x^k} \right] m(\xi)\psi_i(\xi)\, d\xi = 0,$$

$$i = 1,2,3,\ldots, \tag{4.132}$$

is introduced using (4.118) and the system of functions $\psi_i(x)$. Subsequently, series (4.109), (4.122), (4.125), and (4.129) are inserted into (4.132) yielding

$$\int_a^b \sum_k a_k \psi_k(\xi) m(\xi) \psi_i(\xi)\, d\xi$$

$$-\lambda \int_a^b \int_a^b m(x) m(\xi) \left[\sum_r \frac{\psi_r(x)\psi_r(\xi)}{\lambda_r} \right] \left[\sum_l a_l \psi_l(x) \right] \psi_i(\xi)\, dx\, d\xi$$

$$+A \int_a^b \left[\sum_l a_l \psi_l^{(\nu)}(R) \right] \left[\sum_r \frac{\psi_r^{(k)}(R)}{\lambda_r} \psi_r(\xi) \right] m(\xi)\psi_i(\xi)\, d\xi = 0. \tag{4.133}$$

By means of (4.107) this system of equations can be transformed into

$$\sum_j a_j \left(\delta_{ij} - \frac{\lambda}{\lambda_i} \delta_{ij} + A \psi_j^{(\nu)}(R) \frac{\psi_i^{(k)}(R)}{\lambda_i} \right) = 0. \tag{4.134}$$

For the calculation of the eigenvalue λ, the characteristic equation

$$\det \left(\delta_{ij} - \frac{\lambda}{\lambda_i} \delta_{ij} + A \psi_j^{(\nu)}(R) \frac{\psi_i^{(k)}(R)}{\lambda_i} \right) = 0 \tag{4.135}$$

may be used, which follows obviously from (4.134). The systems of equations (4.133) and (4.134) are actually infinite ones. The question is whether it is permissible to substitute them by finite ones, i.e., whether it is permissible to apply the method of reduction. When this question can be answered in the affirmative, then the "extended method of Grammel" is justified, since it consists likewise as in the case of the actual method of Grammel in replacing the infinite system (4.133) by a finite one.

In order to find an answer to this question, the system (4.134) may be brought into the form

$$a_i \left(1 - \frac{\lambda}{\lambda_i} \right) = -A \frac{\psi_i^{(k)}(R)}{\lambda_i} \sum_{j=1}^{j=\infty} a_j \psi_j^{(\nu)}(R). \tag{4.136}$$

It is obvious that the coefficient a_i of the series expansion is obtained as

$$a_i = -A \frac{\psi_i^{(k)}(R)}{\lambda_i - \lambda} \sum_{j=1}^{j=\infty} a_j \psi_j^{(\nu)}(R). \tag{4.137}$$

Using instead the reduced finite system of equations, one would obtain for the same coefficient the approximation

$$\bar{a}_i = -A \frac{\psi_i^{(k)}(R)}{\lambda_i - \lambda} \sum_{j=1}^{j=N} a_j \psi_j^{(\nu)}(R), \tag{4.138}$$

where N has a prescribed value. If N is large enough, the right hand sides of (4.137) and (4.138) differ arbitrarily little from each other since the series $\sum_j a_j \psi_j^{(\nu)}(R)$ is convergent. Consequently, the approximation $\bar{a}_i$ differs arbitrarily little from the exact value a_i. The result is that for the infinite system (4.134) as well as for (4.133) the method of reduction is permissible. Finally, it has to be clarified to which degree of accuracy the eigenvalue λ, the very object of the calculations, may be determined.

Equation (4.132) is exact. Instead of (4.132),

$$\int_a^b \left[y^{(n)}(\xi) - \bar{\lambda} \int_a^b \bar{G}(x,\xi) m(x) y^{(n)}(x) dx \right.$$
$$\left. + A\left(y^{(n)}(R)\right)^{(\nu)} \frac{\partial^k \bar{G}(R,\xi)}{\partial x^k} \right] m(\xi) \psi_i(\xi) d\xi = 0 \tag{4.139}$$

is used in the calculations, where $y^{(n)}$ and $(y^{(n)})^{(\nu)}$ are obtained from (4.125) and (4.129) as approximations of y and $y^{(\nu)}$ by replacing the corresponding infinite series by finite ones. $\bar{\lambda}$ is an approximation of λ.

Subtracting (4.132) from (4.139) and adding

$$\int_a^b \left[\lambda \int_a^b \bar{G}(x,\xi) m(x) y^{(n)}(x) dx - \lambda \int_a^b \bar{G}(x,\xi) m(x) y^{(n)}(x) dx \right] m(\xi) \psi_i(\xi) d\xi,$$

by which nothing is being changed, one arrives after some simple calculations at

$$(\bar{\lambda} - \lambda) \int_a^b \int_a^b \bar{G}(x,\xi) m(x) m(\xi) y(x) \psi_i(\xi) \, dx \, d\xi$$
$$= \int_a^b [y^{(n)}(\xi) - y(\xi)] m(\xi) \psi_i(\xi) d\xi$$

$$-\lambda \int_a^b \int_a^b [y^{(n)}(x) - y(x)] \bar{G}(x, \xi) m(x) m(\xi) \psi_i(\xi) \, dx \, d\xi$$
$$+A \, [(y^{(n)}(R))^{(\nu)} - y^{(\nu)}(R)] \int_a^b \frac{\partial^k \bar{G}(R, \xi)}{\partial x^k} m(\xi) \psi_i(\xi) \, d\xi. \quad (4.140)$$

By virtue of (4.107), (4.109) and (4.122), equation (4.140) can be transformed into

$$(\bar{\lambda} - \lambda) \int_a^b m(x) y^{(n)}(x) \psi_i(x) \, dx$$
$$= \lambda_i \int_a^b [y^{(n)}(\xi) - y(\xi)] m(\xi) \psi_i(\xi) \, d\xi$$
$$-\lambda \int_a^b [y^{(n)}(x) - y(x)] m(x) \psi_i(x) \, dx \quad (4.140)$$
$$+A \, [(y^{(n)}(R))^{(\nu)} - y^{(\nu)}(R)] \, \psi_i^{(k)}(R).$$

This relationship can be changed further into

$$(\bar{\lambda} - \lambda) \int_a^b m(x) y^{(n)}(x) \psi_i(x) \, dx$$
$$= (\lambda_i - \lambda) \int_a^b [y^{(n)}(t) - y(t)] m(t) \psi_i(t) \, dt \quad (4.142)$$
$$+A \, [(y^{(n)}(R))^{(\nu)} - y^{(\nu)}(R)] \, \psi_i^{(k)}(R)$$

by renaming the variables of integration.

However,

$$\int_a^b m(x) y^{(n)}(x) \psi_i(x) \, dx = a_i$$

is a finite quantity due to the convergence of (4.125). The first term on the right hand side of (4.142) can be estimated by means of the inequality of Schwarz yielding

$$\left| \int_a^b [y^{(n)}(t) - y(t)] m(t) \psi_i(t)\, dt \right| \leqq M \left[\int_a^b [y^{(n)}(t) - y(t)]^2\, dt \right]^{1/2}, \quad (4.143)$$

where M is the mean value of m. Using these results, one finds that

$$|\bar{\lambda} - \lambda| \leqq \left| \frac{\lambda_i - \lambda}{a_i} \right| M \left[\int_a^b [y^{(n)}(t) - y(t)]^2\, dt \right]^{1/2} + \left| \frac{A\psi_i^{(k)}(R)}{a_i} \right| \left| \left(y^{(n)}(R)\right)^{(\nu)} - y^{(\nu)}(R) \right| . \quad (4.144)$$

Considering the convergence of the series (4.125) and (4.129), one can conclude that

$$\left[\int_a^b [y^{(n)}(t) - y(t)]^2\, dt \right]^{1/2} < \epsilon, \left| \left(y^{(n)}(R)\right)^{(\nu)} - y^{(\nu)}(R) \right| < \epsilon \quad (4.145)$$

for sufficiently large n, where ϵ is an arbitrary small, positive number. Hence,

$$|\bar{\lambda} - \lambda| \leqq \left(\left| \frac{\lambda_i - \lambda}{a_i} \right| + \left| \frac{A\psi_i^{(k)}(R)}{a_i} \right| \right) \epsilon \quad (4.146)$$

follows from (4.144). This result implies the convergence of the approximate value $\bar{\lambda}$ toward the exact value λ, since $\epsilon \to 0$ for $n \to \infty$, and all the other quantities on the right hand side of (4.146) are bounded.

As an example, the problem

$$y'' = -\lambda y \quad (4.147)$$

$$y(0) = 0, \quad y(1) + y'(1) = 0 \quad (4.148)$$

may be solved by means of the extended method of Grammel. The numerical calculations have been carried out by Dr. K. Nixdorff.

The first eigenvalue of the problem is $\lambda = 4.11585$.

As a substitute problem,

$$y'' = 0, \quad (4.149)$$

$$y(0) = 0, \quad y'(1) = 0 \quad (4.150)$$

may be chosen, whose Green function reads

$$\bar{G}(x,\xi) = \begin{cases} x \text{ for } x \leqq \xi, \\ \xi \text{ for } x \geqq \xi. \end{cases} \tag{4.151}$$

By means of Green's identity

$$\int_0^1 \lambda \bar{G}(x,\xi) y(x)\,dx - y(\xi)$$
$$= \left[-\bar{G}(1,\xi)y'(1) + y(1)\frac{\partial \bar{G}}{\partial x}(1,\xi) + \bar{G}(0,\xi)y'(0) - y(0)\frac{\partial \bar{G}}{\partial x}(0,\xi) \right],$$

and taking the boundary conditions (4.148) for y and (4.150) for $\bar{G}(x, \xi)$ into account, one arrives at

$$y(\xi) = \lambda \int_0^1 \bar{G}(x,\xi) y(x)\,dx + y'(1)\bar{G}(1,\xi). \tag{4.152}$$

Because of the choice of the substitute problem (4.149), (4.150) and the second condition in (4.148), the relationship $y'(1) = -y(1)$ must be used in (4.152), thus, yielding

$$y(\xi) = \lambda \int_0^1 \bar{G}(x,\xi) y(x)\,dx - y(1)\bar{G}(1,\xi). \tag{4.153}$$

This equation corresponds to (4.117). It is obvious that all conditions for convergence are met. Since $\mu = 2$, $\mu/2-1 = 0$. In the weighing term of (4.153), i.e., $y(1)\,\bar{G}(1, \xi)$, do in fact only y and $\bar{G}$ themselves appear. That means, only derivatives of zero order occur in the weighing term, as it should be the case. In addition, $\bar{G}(x, \xi)$ is bilinearly expandable by means of

$$\bar{G}(x,\xi) = \begin{Bmatrix} x \\ \xi \end{Bmatrix} = \frac{2}{\pi^2} \sum_{k=1}^{\infty} \frac{\sin(k-1/2)\pi x \sin(k-1/2)\pi\xi}{(n-1/2)^2} \text{ for } \begin{Bmatrix} x \leqq \xi \\ x \geqq \xi \end{Bmatrix}. \tag{4.154}$$

In the sense of the extended method of Grammel, and corresponding to (4.132), the system of equations

$$\int_0^1 \left[y(\xi) - \lambda \int_0^1 \bar{G}(x,\xi) y(x)\,dx + y(1)\,\bar{G}(1,\xi) \right] \psi_i(\xi)\,d\xi = 0 \tag{4.155}$$

is introduced. As coordinate functions, the functions

$$\psi_i(x) = \sin\left(i - \frac{1}{2}\right)\pi x \tag{4.156}$$

are chosen. They satisfy the boundary conditions $\psi_i(0)$, $\psi_i'(1) = 0$ of the substitute problem.

The extended equations of Grammel read

$$\sum_k a_k \left[\int_0^1 \{\psi_k(\xi) - \lambda \int_0^1 \overline{G}(x,\xi)\,\psi_k(x)dx + \psi_k(1)\xi\}\psi_i(\xi)d\xi \right] = 0, \tag{4.157}$$

where $\overline{G}(1,\xi) = \xi$ and $y = \sum_k a_k\,\psi_k$ have been used. These equations correspond to the system (4.133). The system (4.157) is being solved by means of reducing it into a finite system. In this manner, $\overline{\lambda}$ is obtained as an approximation of λ from

$$\det\left[\int_0^1 \{\psi_k(\xi) - \overline{\lambda} \int_0^1 \overline{G}(x,\xi)\,\psi_k(x)dx + \psi_k(1)\xi\}\psi_i(\xi)d\xi \right] = 0, \tag{4.158}$$

$i, k = 1, 2, \ldots, N.$

The result is

$\overline{\lambda} = 4.26680;$ 4.18600; 4.14262; 4.11560;

for $N =$ 2; 3; 4; 5.

The convergence of $\overline{\lambda}$ toward the exact value $\lambda = 4.11585$ for an increasing number N of coordinate functions is clearly visible.

Instead of (4.156), the system of eigenfunctions of $\overline{G}(x,\xi)$ (see (4.154)), the more general system

$$\psi_i(x) = x^{i+1} - \frac{i+1}{i}\,x^i \tag{4.159}$$

of coordinate functions may be used, which satisfies likewise the boundary conditions (4.150) of the substitute problem. Then, the following approximations

$\overline{\lambda} = 4.29744;$ 4.20935; 4.17725; 4.15991;

for $N =$ 2; 3; 4; 5;

are obtained from (4.158) by means of an analogous sequence of operations as before.

Finally, one might believe that the substitute problem

$$y'' = 0, \tag{4.160}$$

$$y(0), \quad y(1) = 0, \tag{4.161}$$

$$\bar{\bar{G}}(x,\xi) = \begin{cases} x(1-\xi) \text{ for } x \leqq \xi, \\ (1-\xi) \text{ for } x \geqq \xi \end{cases} \tag{4.162}$$

were also suitable, since $\bar{\bar{G}}$ is again bilinearly expandable. Using Green's identity and the boundary conditions (4.161) for $\bar{\bar{G}}$ as well as the boundary conditions (4.148) for y, one arrives at

$$y(\xi) = \lambda \int_0^1 \bar{\bar{G}}(x,\xi) y(x) dx + y'(1) \frac{\partial \bar{\bar{G}}(1,\xi)}{\partial x}, \tag{4.163}$$

which corresponds to equation (4.117). For the derivation of (4.163), condition $y(1) = -y'(1)$ has been used, which follows from the second condition in (4.148) and the choice of (4.160), (4.161) as the substitute problem. However, one notices that the weighing term involves now the quantities $y'(1)$ and $\partial\bar{\bar{G}}(1, \xi)\, \partial x$. This is not permissible, since only derivatives of the order $\mu/2-1 = 2/2-1 = 0$ are allowed. A numerical calculation proves fast, that the method does not converge and yields useless approximations of λ when carried out based on the choice of (4.162) and (4.163).

4.4 Method of Trefftz

Trefftz himself has called his method "a counterpart to the method of Ritz". As will be shown here, it is better to consider it as being reciprocal to the method of Galerkin.

The fundamental variational principle (1.19) can be written in the form

$$\int_T \left[\int_B \left\{ \frac{d}{dt}\left(\frac{\partial \mathscr{T}}{\partial w_t}\right) - (Q_t[w] + Q_x[w]) \right\} \delta w dB + \delta U - \int_R P[w] \delta w dR \right] dt = 0, \tag{4.165}$$

using (1.39)

Now, assume the supported part R_L of the boundary released from its constraints. Then, the total potential energy U has to be replaced by

$$U^* = U - \int_{R_L} \bar{\mathscr{F}}_1(w)\, \bar{\mathscr{D}}_1(w) dR = \int_B \mathscr{U} dB - \int_{R_L} \bar{\mathscr{F}}_1(w)\, \bar{\mathscr{D}}_1(w) dR. \tag{4.166}$$

In (4.166), $\bar{\mathscr{F}}_1(w)\bar{\mathscr{D}}_1(w)$ is the density of the work done along R_L if the boundary is freed from its support. $\bar{\mathscr{F}}_1(w)$ is that "vector" of *those* internal forces which are generated as "external" forces along R_L as a consequence of releasing this part of the boundary from its constraints. The quantity $\bar{\mathscr{D}}_1(w)$ is the "vector" of all displacements (in the general sense), which correspond to the internal forces having been made external ones. The negative sign in front of the line (or surface) integral in (4.166) indicates that one has to deal with a "potential" as U is one. Inserting

$$\delta U^* = \int_B \delta \mathscr{U} dB - \int_{R_L} \bar{\mathscr{F}}_1(\delta w)\bar{\mathscr{D}}_1(w)dR - \int_{R_L} \bar{\mathscr{F}}_1(w)\bar{\mathscr{D}}_1(\delta w)dR \tag{4.167}$$

instead of

$$\delta U = \int_B \delta \mathscr{U} dB$$

in (4.165), the *extended fundamental variational principle*

$$\int_T \Bigg[\int_B \frac{d}{dt}\frac{\partial \mathscr{T}}{\partial w_t}\delta w + \delta \mathscr{U} - (Q_t[w] + Q_x[w])\delta w \; dB$$

$$- \int_{R_L} \bar{\mathscr{F}}_1(\delta w)\bar{\mathscr{D}}_1(w)dR - \int_{R_L} \bar{\mathscr{F}}_1(w)\bar{\mathscr{D}}_1(\delta w)dR - \int_R P[w]\delta w dR \Bigg] dt = 0 \tag{4.168}$$

is obtained after a simple rearrangement. By means of the transformations discussed in Section 1.2, the relationship

$$\int_T\int_B \delta \mathscr{U} dB\, dt = \int_T \Bigg[\int_B \frac{\delta U}{\delta w}\delta w dB + \int_R W(w, \delta w)dR \Bigg] dt \tag{1.45}$$

can be set up, which is generally valid, therefore, also here. According to the here introduced notations,

$$\int_R W(w, \delta w)dR \equiv \int_R \bar{\mathscr{F}}(w)\bar{\mathscr{D}}(\delta w)dR, \tag{4.169}$$

where $\bar{\mathscr{F}}(w)$ is the vector of all possible internal forces and $\bar{\mathscr{D}}(\delta w)$ the vector of the corresponding virtual displacements. As a consequence of (4.169), equation (1.45) changes into

$$\int_T\int_B \delta\mathscr{U}\, dB\, dt = \int_T\left[\int_B \frac{\delta U}{\delta w}\,\delta w\, dB + \int_R \bar{\mathscr{F}}(w)\,\bar{\mathscr{D}}(\delta w)\, dR\right] dt. \qquad (4.170)$$

Using (4.170) in (4.168), and taking into account that

$$R = R_L + R_F, \qquad (4.171)$$

yields the *extended derived variational principle*

$$\int_T\Bigg[\int_B \left\{\frac{d}{dt}\left(\frac{\partial\mathscr{T}}{\partial w_t}\right) + \frac{\delta U}{\delta w} - (Q_t[w] + Q_x[w])\right\}\delta w\, dB$$

$$+ \int_{R_L} [\bar{\mathscr{F}}(w)\bar{\mathscr{D}}(\delta w) - \bar{\mathscr{F}}_1(w)\bar{\mathscr{D}}_1(\delta w) - \bar{\mathscr{F}}_1(\delta w)\bar{\mathscr{D}}_1(w)$$

$$- P[w]\delta w] dR + \int_{R_F} [\bar{\mathscr{F}}(w)\bar{\mathscr{D}}(\delta w) - P[w]\delta w] dR\Bigg] dt = 0. \qquad (4.172)$$

This principle will be used as the basis for all considerations to follow.

Now, $Q_t[w] \equiv 0$ shall be assumed here and subsequently. Moreover, under the assumption that

$$w = f(t)y(x_1, x_2), \qquad (4.173)$$

also

$$Q_x[w] = fQ_x[y], \quad P[w] = fP[y]. \qquad (4.174)$$

Furthermore,

$$\frac{d}{dt}\left(\frac{\partial\mathscr{T}}{\partial w_t}\right) = yTf, \qquad (4.175)$$

$$\frac{\delta U}{\delta w} = f\,\frac{\delta U}{\delta y}, \qquad (4.176)$$

$$\delta w = f\delta y, \qquad (4.177)$$

and

$$\frac{Tf}{f} = -\lambda \tag{1.101}$$

shall be used. Then, (4.172) changes after all this, after suppressing $\int_T f^2 dt$, and taking (4.171) into consideration, into

$$\int_B \left\{ -\lambda y + \frac{\delta U}{\delta y} + Q_x [y] \right\} \delta y \, dB + \int_{R_L} \Big[\bar{\mathscr{F}}(y) \bar{\mathscr{D}}(\delta y) - \bar{\mathscr{F}}_1(y) \bar{\mathscr{D}}_1(\delta y)$$

$$- \bar{\mathscr{F}}_1(\delta y) \bar{\mathscr{D}}_1(y) - P[y] \delta y \Big] dR + \int_{R_F} \Big[\bar{\mathscr{F}}(y) \bar{\mathscr{D}}(\delta y) - P[y] \delta y \Big] dR = 0. \tag{4.178}$$

The *classical method* in the calculus of variations consists in assuming in (4.178) that δy be arbitrary in B. In addition it may be assumed that due to the boundary being completely released from its constraints (a fact which has been considered by substituting U by U^*),

δw is arbitrary on R_L and R_F.

Consequently,

$$-\lambda y + \frac{\delta U}{\delta y} + Q_x [y] = 0, \tag{4.179}$$

$$\bar{\mathscr{D}}_1 (y) = 0 \text{ on } R_L, \tag{4.180}$$

$$\begin{aligned} &\bar{\mathscr{F}}(y) - \bar{\mathscr{F}}_1(y) - P[y] = 0 \text{ componentwise on } R_L, \\ &\bar{\mathscr{F}}(y) - P[y] \qquad\qquad = 0 \text{ componentwise on } R_F \end{aligned} \tag{4.181}$$

follows from (4.178). The differential equation of the problem is (4.179), and (4.180), (4.181) are the geometric and dynamic boundary conditions of the problem.

The *direct method* in the calculus of variations consists in replacing y by an approximation y^* in (4.178), and in demanding that

$$\int_B \left\{ -\lambda y^* + \frac{\delta U}{\delta y^*} + Q_x [y^*] \right\} \delta y^* \, dB$$

$$+ \int_{R_L} \Big[\bar{\mathscr{F}}(y^*) \bar{\mathscr{D}}(\delta y^*) \bar{\mathscr{F}}_1(\delta y^*) - \bar{\mathscr{D}}_1(\delta y^*) - \bar{\mathscr{F}}_1(\delta y^*) \bar{\mathscr{D}}_1(y^*)$$

$$-P[y^*]\delta y^*\Bigg] dR + \int_{R_F} \Bigg[\bar{\mathscr{F}}(y^*)\bar{\mathscr{D}}(\delta y^*) - P[y^*]\delta y^* \Bigg] dR = 0 \quad (4.182)$$

holds true at least approximately. Subsequently, one may follow two paths:

(i) The *Galerkin way* consists in assuming that y^* satisfies all boundary conditions so that

$$\int_{R_L} \Bigg[\bar{\mathscr{F}}(y^*)\bar{\mathscr{D}}(\delta y^*) - \bar{\mathscr{F}}_1(y^*)\bar{\mathscr{D}}_1(\delta y^*) - \bar{\mathscr{F}}_1(\delta y^*)\bar{\mathscr{D}}_1(y^*)$$

$$-P[y^*]\delta y^*\Bigg] dR \equiv 0. \quad (4.183)$$

Because of (4.183),

$$\int_B \left\{ -\lambda y^* + \frac{\delta U}{\delta y^*} + Q_x[y^*] \right\} \delta y^* \, dB = 0 \quad (4.184)$$

remains from (4.182) to be used for determining y^*. Using in addition $y^* = \sum_i a_i y_i$, where the coordinate functions y_i satisfy the boundary conditions, one obtains the *equations of Galerkin*.

(ii) The reciprocal *way of Trefftz* consists in assuming that y^* satisfies the differential equation so that

$$\int_B \left\{ -\lambda y^* + \frac{\delta U}{\delta y^*} + Q_x[y^*] \right\} \delta y^* \, dB \equiv 0, \quad (4.185)$$

since the content of the curled bracked in (4.185) is identical zero. Therefore,

$$\int_{R_L} \Bigg[\bar{\mathscr{F}}(y^*)\bar{\mathscr{D}}(\delta y^*) - \bar{\mathscr{F}}_1(y^*)\bar{\mathscr{D}}_1(\delta y^*) - \bar{\mathscr{F}}_1(\delta y^*)\bar{\mathscr{D}}_1(y^*)$$

$$-P[y^*]\delta y^*]dR + \int_{R_F} [\bar{\mathscr{F}}(y^*)\bar{\mathscr{D}}(\delta y^*) - P[y^*]\delta y^*\Bigg] dR = 0 \quad (4.186)$$

remains from (4.182) as an equation for determining y^*. Now, assume

$$y^* = y_0 + \sum_i a_i y_i, \quad (4.187)$$

where y_0 is a particular solution of the differential equation (4.179), and where the coordinate functions y_i satisfy the *homogeneous* differential equation corresponding to (4.179). Using (4.187) in (4.186) yields the *equations of Trefftz*. An application of these equations shall be shown by means of two examples.

The first example shall be a shaft under torsion. Its length be l, its modulus of rigidity G and its twist θ. The total potential is

$$U = \frac{1}{2G} \int_B (F_{x_1}^2 + F_{x_2}^2 - 4G\theta F)\, dB, \tag{4.188}$$

where F is the torsion function and B the cross-section of the shaft having the contour R: The quantity U^* which is generated by releasing the edge of the cross-section from its constraints reads (according to [47])

$$U^* = U - \int_R F \frac{\partial F}{\partial n}\, dR - \frac{l}{2G} \int_B (F_{x_1}^2 + F_{x_2}^2 - 4G\theta F)\, dB - \frac{l}{G} \int_R F \frac{\partial F}{\partial n}\, dR.$$

Applying Green's formula

$$\int_B u \Delta v\, dB = \int_R u \frac{\partial v}{\partial n}\, dB - \int_B (u_{x_1} v_{x_1} + u_{x_2} v_{x_2})\, dB \tag{4.190}$$

to $u \equiv \delta F$ and $v \equiv F$ yields

$$\delta U^* = -\frac{l}{G} \int_B [\Delta F + 2G\theta]\, \delta F\, dB - \frac{l}{G} \int_R F \frac{\partial \delta F}{\partial n}\, dR. \tag{4.191}$$

Dealing with a static problem like this one, and due to the absence of nonconservative forces, the extended variational principle is reduced to

$$\delta U^* = 0.$$

Therefore, by virtue of (4.191),

$$\int_B [\Delta F + 2G\theta]\, \delta F\, dB + \int_R F \frac{\partial \delta F}{\partial n}\, dR = 0. \tag{4.192}$$

According to the method of Trefftz, an approximation of F in the form

$$F^* = -\frac{G\theta}{2} (x_1^2 + x_2^2) + a_1 + a_2 (x_1^4 - 6x_1^2 x_2^2 + x_2^4) \tag{4.193}$$

is used, where the coordinate functions are

$$\left.\begin{aligned} F_0 &= -\frac{G\theta}{2}(x_1^2 + x_2^2), \\ F_1 &= 1, \\ F_2 &= x_1^4 - 6x_1^2 x_2^2 + x_2^4. \end{aligned}\right\} \tag{4.194}$$

Obviously,

$$\Delta F^* + 2G\theta = 0, \tag{4.195}$$

i.e., F^* satisfies the differential equation of the problem. According to (4.192),

$$\int_R F^* \frac{\partial \delta F^*}{\partial n}\, dR = 0 \tag{4.196}$$

remains for generating the equations of Trefftz. This relationship could have been derived directly from (4.186). It will be shown that this is indeed the case. Corresponding to the nature of this problem,

$$y^* = F^*, R_L \equiv R, R_F \equiv 0, P[F^*] \equiv 0,$$

and according to the *membrane analogy*, the vector of internal forces reads $\bar{\mathscr{F}}(F^*) \equiv \bar{\mathscr{F}}_1(F^*) \equiv (\partial F^*/\partial n)$, and the displacement vector reads $\bar{\mathscr{D}}(F^*) \equiv \bar{\mathscr{D}}_1(F^*) \equiv (F^*)$. With this data, (4.196) is immediately obtained from (4.186). For the assumptions (4.193), (4.194) made for F^*, equation (4.196) becomes

$$a_1 \int_R F_1 \frac{\partial F_2}{\partial n}\, dR + a_2 \int_R F_2 \frac{\partial F_2}{\partial n}\, dR = -\int_R F_0 \frac{\partial F_2}{\partial n}\, dR. \tag{4.197}$$

Assuming a quadratic cross-section with edge length $2a$, and assuming that the origin of the x_1, x_2 – coordinate system is in the middle of the cross-section,

$$\int_R F_1 \frac{\partial F_2}{\partial n}\, dR = 0,$$

$$\int_R F_2 \frac{\partial F_2}{\partial n}\, dR = 32\,\frac{48}{35}\, a^8,$$

$$\int_R F_0 \frac{\partial F_2}{\partial n}\, dR = 32\,\frac{2G\theta}{15}\, a^6$$

is obtained, taking (4.194) into consideration. With these values,

$$a_2 = -\frac{7G\theta}{72a^2} \tag{4.198}$$

follows from (4.197). The coefficient a_1 is determined by demanding that

$$\int_R F^* \, dR = 0, \tag{4.199}$$

thus forcing the function F^* to admit of the boundary value zero, which actually had to be satisfied as close as possible. By means of (4.193), and the value of a_2 given by (4.198),

$$a_1 = \frac{53}{90} G\theta a^2 \tag{4.200}$$

is obtained from (4.199). Inserting (4.198) and (4.200) into (4.193), the approximation F^* of the torsion function F reads

$$F^* = \frac{G\theta}{2a^2}\left[\frac{53}{45}a^4 - a^2(x_1^2 + x_2^2) - \frac{7}{36}(x_1^4 - 6x_1^2x_2^2 + x_2^4)\right] \tag{4.201}$$

An approximation of the torque M_T is obtained by means of (4.201) as

$$M_T^* = 2\int_B F^* \, dB = \frac{304}{135} G\theta a^4 . \tag{4.202}$$

The second example is the quadratic plate with sidelengths $2l$, subjected to the uniformly distributed load p, and clamped along all edges. The plate surface be designated as B and the contour as R. The differential equation of the problem reads

$$\Delta\Delta w = \frac{p}{D}, \tag{4.203}$$

where D is the plate stiffness and w the deflection of the middle surface of the plate.

Releasing the edges from the constraints, edge moments and forces are set free. Since the contour of the plate is piecewise straight, the vector of the internal forces reads

$$\bar{\mathscr{F}}(w) \equiv \bar{\mathscr{F}}_1(w) = \begin{pmatrix} \Delta w \\ -\dfrac{\partial \Delta w}{\partial n} \end{pmatrix}. \tag{4.204}$$

The corresponding displacement vector is

$$\bar{\mathscr{D}}(w) \equiv \bar{\mathscr{D}}_1(w) = \begin{pmatrix} \dfrac{\partial w}{\partial n} \\ w \end{pmatrix} \tag{4.205}$$

The problem is conservative. Therefore, $P[w] \equiv 0$. Moreover, $R_L \equiv R$, and $R_F \equiv 0$. With these data, (4.186) yields

$$\int_R \left[\Delta\delta w^* \frac{\partial w^*}{\partial n} - \frac{\partial \Delta\delta w^*}{\partial n} w^* \right] dR = 0 \tag{4.206}$$

for $y^* \equiv w^*$. This relationship may serve for deriving the equations of Trefftz.

For this purpose, the approximation

$$w^* = \frac{p}{64D}(x_1^2 + x_2^2)^2 + a_1(x_1^2 + x_2^2) \tag{4.207}$$

having the coordinate functions

$$\begin{aligned} w_0 &= \frac{p}{64D}(x_1^2 + x_2^2)^2, \\ w_1 &= x_1^2 + x_2^2 \end{aligned} \tag{4.208}$$

is introduced, which obviously satisfies

$$\Delta\Delta w^* = \frac{p}{D}.$$

By means of (4.207), equation (4.206) changes into

$$\int_R (w_0 + a_1 w_1) \frac{\partial \Delta w_1}{\partial n} dR - \int_R \Delta w_1 \frac{\partial}{\partial n}(w_0 + a_1 w_1)\, dR = 0. \tag{4.209}$$

Using (4.208), the following relationships are obtained

$$\int_R \left(w_1 \frac{\partial \Delta w_1}{\partial n} - \Delta w_1 \frac{\partial w_1}{\partial n} \right) dR = -64 l^2,$$

$$\int_R \left(w_0 \frac{\partial \Delta w_1}{\partial n} - \Delta w_1 \frac{\partial w_0}{\partial n} \right) dR = -8 \frac{p l^4}{3D},$$

which are inserted into (4.209) to yield

$$-8l^2 a_1 - \frac{pl^4}{3D} = 0.$$

Hence,

$$a_1 = -\frac{pl^2}{24D}\,. \tag{4.210}$$

Using (4.210) in (4.207),

$$w^* = \frac{p}{64D}\left[(x_1^2 + x_2^2)^2 - \frac{8}{3}l^2(x_1^2 + x_2^2)\right] \tag{4.211}$$

is obtained as an approximate description of the deflection of the place. However, in order to arrive at

$$w^*(l, l) = 0,$$

which must hold for the deflection of the corner of the plate's edge, the constant

$$k = \frac{pl^2}{48D}$$

has to be added to (4.211), thus yielding finally

$$w^* = \frac{p}{64D}\left[(x_1^2 + x_2^2)^2 - \frac{8}{3}l^2(x_1^2 + x_2^2) + \frac{4l^4}{3}\right]. \tag{4.212}$$

From this expression,

$$w^*(0, 0) = \frac{pl^4}{48D} = 0.0208\,\frac{pl^4}{D}$$

is obtained for the deflection in the middle of the plate. This value compares favourable with the exact value

$$w(0, 0) = 0.0204\,\frac{pl^4}{D}\,.$$

It may not be possible to find an approximation y^* (see page 182) which satisfies the differential equation of the problem. This is in the first place the case when the quantity λ involved does not have a specific value a priori but is an eigenvalue. It may now be assumed that y^* does satisfy only *that* differential equation which is obtained by omitting the disturbing terms. Then, the surface integral over B in (4.182) will not vanish completely or not at all, and the term corresponding to it, which did not vanish, must be added to equation (4.186). If in this equation the quantity y^* is substituted by a sum of coordinate functions, then, in comparison to the equations following from (4.186), one is led this time to the so called *Quasi-Trefftz' equations* due to the additional term mentioned above.

In order to make its application clear, the plate problem shown in Figure 5 may be considered for $a = b = l$ and $\nu = 0.3$. It is then a square plate which is freely supported along R_L, i.e., along $x_1 = 0, x_2 = 0, x_2 = l$, and which is free along R_F, i.e., along $x_1 = l$. This plate is subjected to the compressive *follower* forces N, which are uniformly distributed along the edges $x_1 = 0$, and $x_1 = l$. The problem is to determine the value N_{crit} of N, for which the plate becomes unstable.

According to the data of the problem,

$$Q_x[y] \equiv 0, \quad P[y] = \begin{cases} -Ny_{x_1} & \text{along } R_F, \\ 0 & \text{along } R_L, \end{cases}$$

$$\mathscr{U} = \frac{D}{2}\left[(y_{x_1x_1} + y_{x_2x_2})^2 + 2(1-\nu)(y^2_{x_1x_2} - y_{x_1x_1}y_{x_2x_2}) - \frac{N}{D}y^2_{x_1}\right],$$

$$\bar{\mathscr{F}}_1(y) = \begin{pmatrix}\phi_1 - p\\ 0\end{pmatrix}, \quad \bar{\mathscr{D}}_1(y) = \begin{pmatrix} y\\ 0\end{pmatrix} \text{ along } R_L, \tag{4.213}$$

The "vectors" $\bar{\mathscr{F}}$ and $\bar{\mathscr{D}}$ are given by

$$\bar{\mathscr{F}}(y) = \begin{pmatrix}\phi_1\\ \phi_2\end{pmatrix}, \bar{\mathscr{D}}(y) = \begin{pmatrix} y\\ \partial y/\partial n\end{pmatrix} \text{ along } R = R_L + R_F, \tag{4.214}$$

where the "components" ϕ_1, ϕ_2 read

$$\begin{aligned}\phi_1 = {} & \frac{\partial \mathscr{U}}{\partial y_{x_1}}\cos\alpha + \frac{\partial \mathscr{U}}{\partial y_{x_2}}\sin\alpha - \frac{\partial}{\partial x_1}\left(\frac{\partial \mathscr{U}}{\partial y_{x_1x_1}}\right)\cos\alpha - \frac{\partial}{\partial x_2}\left(\frac{\partial \mathscr{U}}{\partial y_{x_2x_2}}\right)\sin\alpha \\ & - \frac{1}{2}\left[\frac{\partial}{\partial x_1}\left(\frac{\partial \mathscr{U}}{\partial y_{x_1x_2}}\right)\sin\alpha + \frac{\partial}{\partial x_2}\left(\frac{\partial \mathscr{U}}{\partial y_{x_1x_2}}\right)\cos\alpha\right] \\ & + \frac{\partial}{\partial R}\left[\left(\frac{\partial \mathscr{U}}{\partial y_{x_1x_1}} - \frac{\partial \mathscr{U}}{\partial y_{x_2x_2}}\right)\sin\alpha\cos\alpha\right.\end{aligned} \tag{4.215}$$

$$+\frac{1}{2}\left(\frac{\partial \mathcal{U}}{\partial y_{x_1 x_2}}\sin^2\alpha - \frac{\partial \mathcal{U}}{\partial y_{x_1 x_2}}\cos^2\alpha\right)\Bigg],$$

$$\phi_2 = \frac{\partial \mathcal{U}}{\partial y_{x_1 x_1}}\cos^2\alpha + \frac{\partial \mathcal{U}}{\partial y_{x_2 x_2}}\sin^2\alpha + \frac{\partial \mathcal{U}}{\partial y_{x_1 x_2}}\sin\alpha\cos\alpha.$$

The angle α appearing in these formulas is the angle between the exterior normal of the plate's contour and the x_1 direction. Using these relationships, equation (4.182) yields

$$\iint_B \left\{-\lambda y^* + \frac{\delta U}{\delta y^*}\right\}\delta y^* dB + \int_{R_L}\left[\phi_2(y^*)\frac{\partial \delta y^*}{\partial n} - \phi_1(\delta y^*)y^*\right]dR$$

$$+\int_{R_F}\left[\phi_1(y^*)\,\delta y^* + \phi_2(y^*)\frac{\partial \delta y^*}{\partial n} + Ny^*_{x_1}\,\delta y^*\right]dR = 0. \tag{4.216}$$

Using in addition the value of $\mathcal{U}$ given in (4.213), one finds due to (1.42) (here applied to y, y^*, respectively, instead of w), and by means of (4.215) that

$$\frac{\delta U}{\delta y^*} = D\Delta^2 y^* + Ny^*_{x_1 x_1}, \tag{4.217}$$

$$\left.\begin{aligned}\phi_1(y^*) &= -D\frac{\partial \Delta y^*}{\partial n}\\ \phi_2(y^*) &= D\Delta y^*\end{aligned}\right\}\quad \begin{aligned}&\text{along } x_2 = 0 \text{ and } x_2 = l,\\ &\text{i.e., along } R_1,\end{aligned} \tag{4.218}$$

$$\left.\begin{aligned}\phi_1(y^*) &= -D\frac{\partial \Delta y^*}{\partial n}\\ \phi_2(y^*) &= D\Delta y^*\end{aligned}\right\}\quad \begin{aligned}&\text{along } x_1 = 0,\\ &\text{i.e., along } R_2,\end{aligned} \tag{4.219}$$

where $R_1 + R_2 = R_L$. Also, the fact has been used that $y^*_{x_1 x_2} = 0$ along R_1 and $y^*_{x_2 x_2} = 0$ along R_2 due to the conditions of support of the plate.

On the other hand,

$$\left.\begin{aligned}\phi_1(y^*) &= -D\frac{\partial}{\partial n}[y^*_{x_1 x_1} + (2-\nu)y^*_{x_2 x_2} - Ny^*_{x_1}],\\ \phi_2(y^*) &= D(y^*_{x_1 x_1} + \nu y^*_{x_2 x_2})\end{aligned}\right\}\quad \begin{aligned}&\text{along } x_1 = l,\\ &\text{i.e., along } R_F.\end{aligned} \tag{4.220}$$

Using (4.217), (4.218), (4.219), and (4.220) in (4.216) yields

$$\int_B \{-\lambda y^* + D\Delta^2 y^* + N y^*_{x_1 x_1}\}\delta y^* dB + D \int_{R_L} \left(y^* \frac{\partial \Delta y^*}{\partial n} + \Delta y^* \frac{\partial \delta y^*}{\partial n} \right) dR$$

$$+ D \int_{R_F} \left\{ -\delta y^* \frac{\partial}{\partial n} \left[y^*_{x_1 x_1} + (2-\nu) y^*_{x_2 x_2} \right] \right.$$

$$\left. + \frac{\partial \delta y^*}{\partial n} \left[y^*_{x_1 x_1} + \nu y^*_{x_2 x_2} \right] \right\} dR = 0. \qquad (4.221)$$

In order to arrive at the Quasi-Trefftz equations, y^* has to be substituted in (4.221) by

$$y^* = \sum_{i=1}^{n} a_i \phi_i . \qquad (4.222)$$

Assuming that the coordinate functions satisfy neither the differential equation nor the boundary conditions, and considering that the variations δa_i are arbitrary, the system

$$\sum_{i=1}^{n} a_i \left\{ [-\lambda\phi_i + D\Delta^2 \phi_i + N\phi_{ix_1 x_1}]\phi_j dB + D \int_{R_L} \left(\phi_i \frac{\partial \Delta \phi_j}{\partial n} + \Delta\phi_i \frac{\partial \phi_j}{\partial n} \right) dR \right.$$

$$+ D \int_{R_F} \left\{ -\phi_j \frac{\partial}{\partial n} [\phi_{ix_1 x_1} + (2-\nu)\phi_{ix_2 x_2}] \right.$$

$$\left. \left. + \frac{\partial \phi_j}{\partial n} [\phi_{ix_1 x_1} + \nu \phi_{ix_2 x_2}] \right\} dR \right\} = 0, \quad j = 1,2, \ldots, n, \qquad (4.223)$$

is obtained. It is the system of Quasi-Trefftz equations of this specific example.

In order to show the usefulness of the system, it may be applied to the plate problem under consideration using the single coordinate function

$$\phi_i = \phi_j = \phi = x_1 x_2^2 - x_1^3 \frac{x_2}{l} . \qquad (4.224)$$

As one can see easily, it depicts relatively well the first eigenmode of the buckled plate.

As already indicated in Figures 23 a and b in Section 2.3 for the problem in Figure 5, the instability of the plate occurs through buckling and not flutter. Therefore, it may be assumed from the beginning on, that the buckling load is obtained at $\omega^2 = 0$, i.e., at $\lambda = 0$. Moreover, it turns out that $\Delta^2\phi = 0$. Taking all this into account and rewriting (4.223) in terms of ϕ yields

$$\int_B \frac{N}{D}\phi_{x_1x_1}\phi dB + \int_{R_L}\left(\phi\frac{\partial\Delta\phi}{\partial n} + \Delta\phi\frac{\partial\phi}{\partial n}\right)dR \tag{4.225}$$

$$+\int_{R_F}\left\{-\phi\frac{\partial}{\partial n}\left[\phi_{x_1x_1} + (2-\nu)\phi_{x_2x_2}\right] + \frac{\partial\phi}{\partial n}\left[\phi_{x_1x_1} + \nu\phi_{x_2x_2}\right]\right\}dR = 0.$$

Considering the specific circumstances of the plate in question, equation (4.225) can be changed into

$$\frac{N}{D}\int_{x_2=0}^{l}\int_{x_1=0}^{l}\frac{\partial^2\phi}{\partial x_1^2}\phi dx_1 dx_2 + \int_0^l\left(-\phi\frac{\partial\Delta\phi}{\partial x_2} - \Delta\phi\frac{\partial\phi}{\partial x_2}\right)_{x_2=0} dx_1$$

$$-\int_0^l\left(\phi\frac{\partial\Delta\phi}{\partial x_2} + \Delta\phi\frac{\partial\phi}{\partial x_2}\right)_{x_2=l} dx_1 - \int_0^l\left(-\phi\frac{\partial\Delta\phi}{\partial x_1} - \Delta\phi\frac{\partial\phi}{\partial x_1}\right)_{x_1=0} dx_2$$

$$+\int_0^l\left\{-\phi\frac{\partial}{\partial x_1}\left[\frac{\partial^2\phi}{\partial x_1^2} + (2-\nu)\frac{\partial^2\phi}{\partial x_2^2}\right]\right.$$

$$\left.+\frac{\partial\phi}{\partial x_1}\left[\frac{\partial^2\phi}{\partial x_1^2} + \nu\frac{\partial^2\phi}{\partial x_2^2}\right]\right\}_{x_1=l} dx_2 = 0. \tag{4.226}$$

After some calculation one obtains the relationships

$$\frac{\partial\phi}{\partial x_1} = x_2^2 - 3x_1^2\frac{x_2}{l}, \qquad \frac{\partial\phi}{\partial x_2} = 2x_1x_2 - \frac{x_1^3}{l},$$

$$\frac{\partial^2\phi}{\partial x_1^2} = -6x_1\frac{x_2}{l}, \qquad \frac{\partial^2\phi}{\partial x_2^2} = 2x_1,$$

$$\Delta\phi = x_1\left(2 - \frac{6x_2}{l}\right), \tag{4.227}$$

$$\frac{\partial\Delta\phi}{\partial x_1} = 2 - \frac{6x_2}{l}, \qquad \frac{\partial\Delta\phi}{\partial x_2} = -\frac{6x_1}{l}.$$

Using these, $\nu = 0.3$, and (4.224), equation (4.226) yields after integration

$$-\frac{N}{D}\frac{l^6}{10}+\frac{86l^4}{15}=0.$$

Solving this equation appropriately, the first buckling load of the plate is determined approximately as

$$N^*_{\text{crit}}=\frac{860}{15}\frac{D}{l^2}=57.333\frac{D}{l^2}.$$

This value agrees in an acceptable way with the value

$$N_{\text{crit}}=50.72\frac{D}{l^2},$$

which was calculated previously in Section 2.3.

It may be of interest to carry out some considerations about the convergence of the Quasi-Trefftz equations. Convergence of those implies of course the convergence of the actual equations of Trefftz which are only a special case of the more general Quasi-Trefftz equations.

All the proofs of convergence of the method of Trefftz, which can be found in the literature are based on the fact that there exists a functional which has the property to assume a minimum for the true solution of the problem under consideration. But this is the fact for conservative problems only. Therefore, a different proof of convergence has to be found for nonconservative problems which is free of any assumption about extremal properties of the functional used in the method of Trefftz. Deliberations of this kind will be carried out in the following. The previous problem will be used as an illustration. It is general enough to make the special features of the new proof of convergence apparent.

Let Green's identity

$$\int_B (v\Delta^2 u-u\Delta^2)\,dB=\int_R\left(\frac{\partial\Delta u}{\partial n}v-\Delta u\frac{\partial v}{\partial n}-\frac{\partial\Delta v}{\partial n}u+\Delta v\frac{u}{n}\right)dR \tag{4.228}$$

be introduced. First, the choice for u and v be the following:

(i) $u=y,$ (4.229)

where y is a solution of

$$\left.\begin{aligned}&\Delta^2 y+\lambda\phi y=0,\\&\lambda=\frac{N}{D},\qquad \phi y=y_{x_1x_1}\end{aligned}\right\} \tag{4.230}$$

and satisfies the boundary conditions

$$\left.\begin{array}{ll} y = 0, & \Delta y = 0 \text{ along } R_L, \\ \Delta y = (1-\nu) y_{x_2 x_2}, & \dfrac{\partial \Delta y}{\partial n} = -(1-\nu)\dfrac{\partial y_{x_1 x_1}}{\partial R} \text{ along } R_F \end{array}\right\} \tag{4.231}$$

of the preceding plate problem,

(ii) $v = G(x_1, x_2, \xi_1, \xi_2)$, (4.232)

where G is the Green function of the operator Δ^2 for the boundary conditions (4.231). Then, it is true that

"$\Delta^2 G$ behaves like a Dirac function". (4.233)

Using (4.229) and (4.232) in (4.228) and taking into account that y and G satisfy the boundary conditions (4.231), the result of a simple calculation is

$$y = \int_B G \Delta^2 y \, dB - (1-\nu) \int_{R_F} \left(-\frac{\partial y_{x_1 x_2}}{\partial R} G - y_{x_2 x_2} \frac{\partial G}{\partial n} + \frac{\partial G_{x_1 x_2}}{\partial R} y + G_{x_2 x_2} \frac{\partial y}{\partial n} \right) dR. \tag{4.234}$$

However, according to (4.230),

$$\Delta^2 y = -\lambda \phi y, \tag{4.235}$$

therefore,

$$y = -\lambda \int_B G \phi y \, dB - (1-\nu) \int_{R_F} \left(-\frac{\partial y_{x_1 x_2}}{\partial R} G - y_{x_2 x_2} \frac{\partial G}{\partial n} + \frac{\partial G_{x_1 x_1}}{\partial R} y + G_{x_2 x_2} \frac{\partial y}{\partial n} \right) dR \tag{4.236}$$

follows from (4.234). This is an integro-differential equation for the plate problem under consideration.

Now, the choice for u and v in (4.228) be the following:

(i) $u = y^* = \sum_{i=1}^{n} a_i \phi_i$, (4.237)

where the coordinate functions ϕ_i form a *complete* system but do *not* satisfy the boundary conditions (4.231).

(ii) $v = G$ (4.238)

where G is the Green function defined previously.

Using (4.237) and (4.238) in (4.228), and taking the boundary conditions of G into account, yields the identity

$$y^* = -\int_B G\Delta^2 y^* \, dB - \int_{R_L} \left(\Delta y^* \frac{\partial G}{\partial n} + \frac{\partial \Delta G}{\partial n} y^*\right) dR \tag{4.239}$$

$$+ \int_{R_F} \left[\frac{\partial \Delta y^*}{\partial n} G - \Delta y^* \frac{\partial G}{\partial n} + (1-\nu) \frac{\partial G_{x_1 x_2}}{\partial R} y^* + (1-\nu) G_{x_2 x_2} \frac{\partial y^*}{\partial n}\right] dR = 0.$$

Now, let the Quasi-Trefftz equations be considered. For the sake of brevity they may be written down under the assumption $\lambda \equiv 0$. Using the new notation

$$\lambda^* = \frac{N^*_{\text{crit}}}{D}, \tag{4.240}$$

and taking into account that

$$\left.\begin{aligned} &\frac{\partial}{\partial n} [\phi_{ix_1x_1} + (2-\nu)\phi_{ix_2x_2}] \equiv \frac{\partial \Delta \phi_i}{\partial n} + (1-\nu) \frac{\partial \phi_{ix_1x_2}}{\partial R} \text{ along } R_F, \\ &\phi_{ix_1x_1} + \nu\phi_{ix_2x_2} = \Delta\phi_i - (1-\nu)\phi_{ix_2x_2}, \end{aligned}\right\} \tag{4.241}$$

equation (4.223) yields

$$\sum_{i=1}^{n} a_i \left\{ \int_B (\Delta^2 \phi_i + \lambda^* \phi_{ix_1x_1}) \phi_j dB + \int_{R_L} \left(\phi_i \frac{\partial \Delta \phi_j}{\partial n} + \Delta \phi_i \frac{\partial \phi_j}{\partial n}\right) dR \right.$$

$$+ \int_{R_F} \left\{ -\phi_j \left[\frac{\partial \Delta \phi_i}{\partial n} + (1-\nu) \frac{\partial \phi_{ix_1x_2}}{\partial R}\right] \right. \tag{4.242}$$

$$\left. \left. + \frac{\partial \phi_j}{\partial n} [\Delta \phi_i - (1-\nu)\phi_{ix_2x_2}] \right\} dR \right\} = 0.$$

This relationship can be transformed further into

$$\int_B [\Delta^2 (\sum_i a_i\phi_i) + \lambda^* \phi (\sum_i a_i\phi_i)]\phi_j \, dB + \int_{R_L} \left[(\sum_i a_i\phi_i) \frac{\partial \Delta\phi_j}{\partial n} \right.$$

$$\left. + \Delta(\sum_i a_i\phi_i) \frac{\partial \phi_j}{\partial n} \right] dR$$

$$+ \int_{R_F} \left\{ - \left[\frac{\partial \Delta(\sum_i a_i\phi_i)}{\partial n} + (1-\nu) \frac{\partial (\sum_i a_i\phi_{ix_1x_2})}{\partial R} \right] \phi_j \right.$$

$$\left. + [\Delta(\sum_i a_i\phi_i) - (1-\nu)(\sum_i a_i\phi_{ix_2x_2})] \frac{\partial \phi_j}{\partial n} \right\} dR = 0. \tag{4.243}$$

Let

$$u_j = \int_B G\phi_j \, dB \tag{4.244}$$

be the Fourier coefficients of the series expansion

$$G^* = \sum_{j=1}^{n} u_j \phi_j \tag{4.245}$$

which exists for G at least in the sense of the convergence in the mean. Multiplying (4.243) termwise by u_j and then adding over j yields

$$\int_B (\sum_j u_j\phi_j) \Delta^2 (\sum_i a_i\phi_i) dB + \lambda^* \int_B (\sum_j u_j\phi_j) \phi (\sum_i a_i\phi_i) \, dB$$

$$+ \int_{R_L} \left[(\sum_i a_i\phi_i) \frac{\partial}{\partial n} \Delta(\sum_j a_j\phi_j) + \Delta(\sum_i a_i\phi_i) \frac{\partial}{\partial n} (\sum_j a_j\phi_j) \right] dR$$

$$+ \int_{R_F} \left\{ - \left[\frac{\partial \Delta(\sum_i a_i\phi_i)}{\partial n} + (1-\nu) \frac{\partial (\sum_i a_i\phi_{ix_1x_2})}{\partial R} \right] (\sum_j u_j\phi_j) \right. \tag{4.246}$$

$$\left. + [\Delta(\sum_i a_i\phi_i) - (1-\nu)(\sum_i a_i\phi_{ix_2x_2})] \left(\sum_j u_j \frac{\partial \phi_j}{\partial n} \right) \right\} dR = 0.$$

Because of (4.237) and (4.245) this relationship can be rewritten to read

$$\int_B G^* \Delta^2 y^* \, dB + \lambda^* \int_B G^* \phi y^* \, dB$$

$$+ \int_{R_L} \left(y^* \frac{\partial \Delta G^*}{\partial n} + \Delta y^* \frac{\partial G^*}{\partial n} \right) dR \tag{4.247}$$

$$+ \int_{R_F} \left\{ - \left[\frac{\partial \Delta y^*}{\partial n} + (1-\nu) \frac{\partial y^*_{x_1 x_2}}{\partial R} \right] G^* + [\Delta y^* - (1-\nu) y^*_{x_2 x_2}] \frac{\partial G^*}{\partial n} \right\} dR = 0.$$

Hence, it becomes obvious that calculating λ^* by means of the Quasi-Trefftz equations (4.242) is identical to determining λ^* by means of the integro-differential equation (4.247). This fact will be used in the following convergence considerations.

First, (4.247) be transformed by adding identity (4.239) and by using the fact that

$$(1-\nu) \frac{\partial G_{x_1 x_2}}{\partial R} = - \frac{\partial \Delta G}{\partial n}$$

$$\text{along } R_F, \tag{4.248}$$

$$(1-\nu) G_{x_2 x_2} = \Delta G$$

which follows from the boundary conditions (4.231), if these are applied to G. After some calculation one is led to

$$y^* + \lambda^* \int_B G \phi y^* \, dB + \int_B \Delta^2 y^* (G^* - G) \, dB$$

$$+ \int_{R_L} \left[\Delta y^* \left(\frac{\partial G^*}{\partial n} - \frac{\partial G}{\partial n} \right) + y^* \left(\frac{\partial \Delta G^*}{\partial n} - \frac{\partial \Delta G}{\partial n} \right) \right] dR \tag{4.249}$$

$$+ \int_{R_F} \left\{ \frac{\partial \Delta y^*}{\partial n} (G - G^*) + \Delta y^* \left(\frac{\partial G^*}{\partial n} - \frac{\partial G}{\partial n} \right) \right.$$

$$\left. + (1-\nu) \left[- \frac{\partial y^*_{x_1 x_2}}{\partial R} G^* + y^* \frac{\partial G_{x_1 x_2}}{\partial R} + G_{x_2 x_2} \frac{\partial y^*}{\partial n} - y^*_{x_2 x_2} \frac{\partial G^*}{\partial n} \right] \right\} dR = 0.$$

In orde₁ to investigate convergence, (4.249) must be compared with

$$y+\lambda\int_B G\phi y\,dB \tag{4.250}$$

$$+(1-\nu)\int_{R_F}\left[-\frac{\partial y_{x_1x_2}}{\partial R}G+y\frac{\partial G_{x_1x_2}}{\partial R}+G_{x_2x_2}\frac{\partial y}{\partial n}-y_{x_2x_2}\frac{\partial G}{\partial n}\right]dR=0,$$

where (4.250) follows immediately from (4.236). For the sake of brevity, let the following notations be introduced,

$$\mathscr{A}=y,$$
$$\mathscr{B}=\int_B G\phi y\,dB, \tag{4.251}$$

$$\mathscr{C}=(1-\nu)\int_{R_F}\left(-\frac{\partial y_{x_1x_2}}{\partial R}G+y\frac{\partial G_{x_1x_2}}{\partial R}+G_{x_2x_2}\frac{\partial y}{\partial n}-y_{x_2x_2}\frac{\partial G}{\partial n}\right)dR$$

$$\mathscr{A}^*=y^*,$$
$$\mathscr{B}^*=\int_B G^*\phi y^*\,dB,$$

$$\mathscr{C}^*=(1-\nu)\int_{R_F}\left(-\frac{\partial y^*_{x_1x_2}}{\partial R}G^*+y^*\frac{\partial G_{x_1x_2}}{\partial R}\right.$$

$$\left.+G_{x_2x_2}\frac{\partial y^*}{\partial n}\frac{\partial G^*}{\partial n}\right)dR, \tag{4.252}$$

$$\mathscr{D}^*=\int_B \Delta^2 y^*(G^*-G)\,dB$$

$$+\int_{R_L}\left[\Delta y^*\left(\frac{\partial G^*}{\partial n}-\frac{\partial G}{\partial n}\right)+y^*\left(\frac{\partial\Delta G^*}{\partial n}-\frac{\partial\Delta G}{\partial n}\right)\right]dR$$

$$+\int_{R_F}\left[\frac{\partial\Delta y^*}{\partial n}(G-G^*)+\Delta y^*\left(\frac{\partial G^*}{\partial n}-\frac{\partial G}{\partial n}\right)\right]dR.$$

With these notations,

$$\lambda^* = -\frac{\mathcal{A}^* + \mathcal{C}^* + \mathcal{D}^*}{\mathcal{B}^*} \tag{4.253}$$

follows from (4.249), and

$$\lambda = -\frac{\mathcal{A} + \mathcal{C}}{\mathcal{B}} \tag{4.254}$$

follows from (4.250). Consequently, (4.253) and (4.254) yield

$$\begin{aligned}(\lambda-\lambda^*)\,\mathcal{B}\,\mathcal{B}^* = {} & \mathcal{A}^*(\mathcal{B}-\mathcal{B}^*)+\mathcal{B}^*(\mathcal{A}^*-\mathcal{A}) \\ & +\mathcal{B}^*(\mathcal{C}^*-\mathcal{C})+\mathcal{C}^*(\mathcal{B}-\mathcal{B}^*) \\ & +\mathcal{D}^*(\mathcal{B}-\mathcal{B}^*)+\mathcal{B}^*\mathcal{D}^*.\end{aligned} \tag{4.255}$$

Integrating both sides in (4.255) yields

$$\begin{aligned}(\lambda-\lambda^*)\int_B \mathcal{B}\mathcal{B}^*\,dB + \int_B \mathcal{A}^*(\mathcal{B}-\mathcal{B}^*)\,dB + \int_B \mathcal{B}^*(\mathcal{A}^* \\ -\mathcal{A})\,dB + \int_B \mathcal{B}^*(\mathcal{C}^*-\mathcal{C})\,dB + \int_B \mathcal{C}^*(\mathcal{B}-\mathcal{B}^*)\,dB \\ + \int_B \mathcal{D}^*(\mathcal{B}-\mathcal{B}^*)\,dB \int_B \mathcal{B}^*\mathcal{D}^*\,dB.\end{aligned} \tag{4.256}$$

But then also

$$\begin{aligned}|\lambda-\lambda^*|\cdot\left|\int_B \mathcal{B}\mathcal{B}^*\,dB\right| \leqq \left|\int_B \mathcal{A}^*(\mathcal{B}-\mathcal{B}^*)\,dB\right| + \left|\int_B \mathcal{B}^*(\mathcal{A}^* \right. \\ \left. -\mathcal{A})\,dB\right| + \left|\int_B \mathcal{B}^*(\mathcal{C}^*-\mathcal{C})\,dB\right| + \left|\int_B \mathcal{C}^*(\mathcal{B}-\mathcal{B}^*)\,dB\right| \\ + \left|\int_B \mathcal{D}^*(\mathcal{B}-\mathcal{B}^*)\,dB\right| + \left|\int_B \mathcal{B}^*\mathcal{D}^*\,dB\right|.\end{aligned} \tag{4.257}$$

Applying to (4.257) the well known inequality

$$\left|\int_B u\,v\,dB\right| \leqq \|u\|\cdot\|v\|, \tag{4.258}$$

one finds that

$$|\lambda-\lambda^*|\cdot\left|\int_B \mathscr{B}\mathscr{B}^*\,dB\right| \leqq \|\mathscr{A}^*\|\cdot\|\mathscr{B}-\mathscr{B}^*\|+\|\mathscr{B}^*\|\cdot\|\mathscr{A}^*$$

$$-\mathscr{A}\|+\|\mathscr{B}^*\|\cdot\|\mathscr{C}^*-\mathscr{C}\|+\|\cdot\|\mathscr{B}-\mathscr{B}^*\|+\|\mathscr{D}^*\|\cdot\|$$

$$+\|\mathscr{D}^*\|\cdot\|\mathscr{B}-\mathscr{B}^*\|+\|\mathscr{B}^*\|\cdot\|\mathscr{D}^*\|. \tag{4.259}$$

As one may deduct from (4.259), $\lambda^* \to \lambda$ holds true, i.e., the convergence of the method is ensured, if

$$\begin{array}{lll}
\text{(i)} & \left|\int_B \mathscr{B}\mathscr{B}^*\,dB\right| & \text{is bounded,} \\
\text{(ii)} & \|\mathscr{A}^*\| & \text{is bounded,} \\
\text{(iii)} & \|\mathscr{B}^*\| & \text{is bounded,} \\
\text{(iv)} & \|\mathscr{C}^*\| & \text{is bounded,} \\
\text{(v)} & \|\mathscr{A}^*-\mathscr{A}\| & \text{tends to zero,} \\
\text{(vi)} & \|\mathscr{B}^*-\mathscr{B}\| & \text{tends to zero,} \\
\text{(vii)} & \|\mathscr{C}^*-\mathscr{C}\| & \text{tends to zero,} \\
\text{(viii)} & \|\mathscr{D}^*\| & \text{tends to zero.}
\end{array} \tag{4.260}$$

It will now be shown that conditions (4.260) are satisfied, if the following assumptions hold true:

$$\begin{array}{ll}
(1) & y^*=0 \text{ along } R_2\text{, as a consequence of the chosen coordinate function,} \\
(2) & \int_R y^{*2}\,dR < C_1, \\
(3) & \int_B y^{*2}\,dB < C_2, \\
(4) & \int_{R_F} (y^*_{x_2x_2})^2\,dR < C_3, \\
(5) & \int_B (\phi y^*)^2\,dB < C_4, \\
(6) & \int_R (\Delta y^*)^2\,dR < C_5,
\end{array} \tag{4.261}$$

$$
\begin{aligned}
&(7) \quad \int_{R_F} \frac{\partial y^*_{x_1x_2}}{\partial R}\, dR < C_6,\\
&(8) \quad \int_{R_F} \left(\frac{\partial y^*}{\partial n}\right)^2 dR < C_7,\\
&(9) \quad \int_{R_F} \left(\frac{\partial \Delta y^*}{\partial n}\right)^2 dR < C_8,\\
&(10) \quad \int_{B} (\Delta^2 y^*)^2\, dB < C_9,\\
&(11) \quad \int_{B}\int_{B} G^{*2}\, dB\, dB < C_{10},\\
&(12) \quad \int_{B}\int_{R_F} G^{*2}\, dB\, dR < C_{11},\\
&(13) \quad \int_{B}\int_{R_F} \left(\frac{\partial G^*}{\partial n}\right)^2 dB\, dR < C_{12},\\
&(14) \quad \int_{B} (\phi y)^2\, dB < C_{13},\\
&(15) \quad \int_{R_F} \left(\frac{\partial y}{\partial n}\right)^2 dR < C_{14},\\
&(16) \quad \int_{R_F} (y_{x_2x_2})^2\, dR < C_{15},\\
&(17) \quad \int_{R_F} \left(\frac{\partial y_{x_1x_2}}{\partial R}\right)^2 dR < C_{16},\\
&(18) \quad \int_{B}\int_{B} G^2\, dB\, dB < C_{17},\\
&(19) \quad \int_{B}\int_{R_F} G^2\, dB\, dR < C_{18},\\
&(20) \quad \int_{B}\int_{R_F} G^2_{x_2x_2}\, dB\, dR < C_{19},
\end{aligned}
\qquad (4.261)
$$

$$
\begin{aligned}
&(21)\quad \iint\limits_{BR_F} \left(\frac{\partial G}{\partial n}\right)^2 dB\,dR < C_{20}, \\
&(22)\quad \iint\limits_{BB} \left(\frac{\partial G}{\partial n}\right)^2 dB\,dB < C_{21}, \\
&(23)\quad \iint\limits_{BR_F} \left(\frac{\partial G_{x_1x_2}}{\partial R}\right)^2 dB\,dR < C_{22}, \\
&(24)\quad \int\limits_{B} (y^* - y)^2 dB < \epsilon_1, \\
&(25)\quad \int\limits_{R_F} (y^* - y)^2 dR < \epsilon_2, \\
&(26)\quad \int\limits_{R_F} (y^*_{x_2x_2} - y_{x_2x_2})^2 dR < \epsilon_3, \\
&(27)\quad \int\limits_{R_F} \left(\frac{\partial y^*_{x_1x_1}}{\partial R} - \frac{\partial y_{x_1x_2}}{\partial R}\right)^2 dR < \epsilon_4, \\
&(28)\quad \int\limits_{R_F} \left(\frac{\partial y^*}{\partial n} - \frac{\partial y}{\partial n}\right)^2 dR < \epsilon_5, \qquad (4.261) \\
&(29)\quad \int\limits_{B} (\phi y^* - \phi y)^2 dB < \epsilon_6, \\
&(30)\quad \iint\limits_{BB} (G^* - G)^2 dB\,dB < \epsilon_7, \\
&(31)\quad \iint\limits_{BR_F} (G^* - G)^2 dB\,dR < \epsilon_8, \\
&(32)\quad \int\limits_{BR} \left(\frac{\partial G^*}{\partial n} - \frac{\partial G}{\partial n}\right)^2 dB\,dR < \epsilon_9, \\
&(33)\quad \iint\limits_{BR_L} \left(\frac{\partial \Delta G^*}{\partial n} - \frac{\partial \Delta G}{\partial n}\right)^2 dB\,dR < \epsilon_{10}.
\end{aligned}
$$

C_1 to C_{22} are suitable constants, i.e., it is assumed that the quantities in (4.261,2) to (4.261,23) are square integrable. Likewise ϵ_1 to ϵ_{10} are arbitrary small, positive constants, i.e., it is assumed that the quantities marked by an asterik in

(4.261,24) to (4.261,33) converge in the mean toward the quantities not having an asterik.

Conditions (4.261,1) to (4.261,13) and (4.261,24) to (4.261,33) can always be satisfied by a proper choice of the coordinate functions. As has been shown in [11], it is possible to find a complete system of functions for which convergence in the mean is ensured including their derivatives of prescribed order. Conditions (4.261,14) to (4.261,23) may be satisfied or not, depending on the nature of the problem under discussion. This has to be checked from case to case.

Now, conditions (4.260,i) to (4.260,viii) may be investigated consecutively. During this process, the inequality of Schwarz will be used frequently, sometimes without mentioned especially:

(4.260, i),

$$\left|\int_B \mathscr{B}\mathscr{B}^* dB\right| \leqq \left\{\int_B\left[\int_B G\phi y\, dB\right]^2 dB\right\}^{1/2} \cdot \left\{\int_B\left[\int_B G^*\phi y^*\, dB\right]^2 dB\right\}^{1/2},$$

according to (4.251), (4.252) and (4.258). Using in addition the inequality of Schwarz and the assumptions (4.261,14), (4.261,18), (4.261,5), and (4.261,11),

$$\left|\int_B \mathscr{B}\mathscr{B}^* dB\right| \leqq \left\{\int_B (\phi y)^2 dB \cdot \int_B\int_B G^2 dB\, dB \cdot \int_B (\phi y^*)^2 dB \cdot \int_B\int_B G^{*2} dB\, dB\right\}^{1/2}$$

$$\leqq \left\{C_{13} \cdot C_{17} \cdot C_4 \cdot C_{10}\right\}^{1/2}$$

is obtained, which proves that $\left|\int_B \mathscr{B}\mathscr{B}^* dB\right|$ is bounded.

(4.260,ii):

$$\|\mathscr{A}^*\| = \{\int_B y^{*2} dB\}^{1/2} < C_2^{1/2}$$

according to (4.261,3), i.e., $\|\mathscr{A}^*\|$ is bounded

(4.260,iii):

$$\|\mathscr{B}^*\| = \left\{\int_B\left[\int_B G^*\phi y^*\, dB\right]^{1/2}\right\}$$

$$\leqq \left\{\int_B (\phi y^*)^2 dB \cdot \int_B\int_B G^{*2} dB\, dB\right\}^{1/2} < \left\{C_4 \cdot C_{10}\right\}^{1/2}$$

according to (4.261,5) and (4.261,11) i.e., $\| \mathscr{B}^* \|$ is bounded.
(4.260,iv):

$$\| \mathscr{C}^* \| = \left\{ \int_B \left((1-\nu) \int_{R_F} \left[-\frac{\partial y^*_{x_1x_2}}{\partial R} G^* + y^* \frac{\partial G_{x_1x_2}}{\partial R} \right.\right.\right.$$

$$\left.\left.\left. + G_{x_2x_2} \frac{\partial y^*}{\partial n} - y^*_{x_2x_2} \frac{\partial G^*}{\partial n} \right] dR \right)^2 dB \right\}^{1/2}$$

$$\leqq 2(1-\nu) \left\{ \iint_{B\,R_F} \left[\left(\frac{\partial y^*_{x_1x_2}}{\partial R} \right)^2 G^{*2} + y^{*2} \left(\frac{\partial G_{x_1x_2}}{\partial R} \right)^2 + G^2_{x_2x_2} \left(\frac{\partial y^*}{\partial n} \right)^2 \right.\right.$$

$$\left.\left. + y^{*2}_{x_2x_2} \left(\frac{\partial G^*}{\partial n} \right)^2 \right] dR\, dB \right\}^{1/2}$$

$$\leqq 2(1-\nu) \left\{ \int_{R_F} \left(\frac{\partial y^*_{x_1x_2}}{\partial R} \right)^2 dR \iint_{B\,R_F} G^{*2}\, dB\, dR \right.$$

$$+ \int_{R_F} y^{*2}\, dR \iint_{B\,R_F} \left(\frac{\partial G_{x_1x_2}}{\partial R} \right)^2 dB\, dR + \int_{R_F} \left(\frac{\partial y^*}{\partial n} \right)^2 dR \iint_{B\,R_F} G^2_{x_2x_2}\, dR\, dB$$

$$\left. + \int_{R_F} (y^*_{x_2x_2})^2\, dR \iint_{B\,R_F} \left(\frac{\partial G^*}{\partial n} \right)^2 dB\, dR \right\}^{1/2}$$

$$< 2 \{ C_6 \cdot C_{11} + C_1 \cdot C_{22} + C_7 \cdot C_{19} + C_3 \cdot C_{12} \}^{1/2}$$

according to (4.261,7), (4.261,12), (4.261,2), (4.261,23), (4.261,8), (4.261,20), (4.261,4), and (4.261,13), i.e., $\| \mathscr{C}^* \|$ is bounded.
(4.260,v):

$$\| \mathscr{A}^* - \mathscr{A} \| = \left\{ \int_B (y^* - y)^2\, dB \right\}^{1/2} < \epsilon_1^{1/2}$$

according to (4.261,24), i.e., $\| \mathscr{A}^* - \mathscr{A} \|$ tends toward zero.

$$\|\mathscr{B} - \mathscr{B}^*\| = \left\{ \int_B \left[\int_B G\phi y \, dB - \int_B G^* \phi y^* \, dB \right]^2 dB \right\}^{1/2}$$

$$= \left\{ \int_B \left[\int_B [\phi y (G - G^*) + G^* (\phi y - \phi y^*)] dB \right]^2 dB \right\}^{1/2}$$

$$\leqq \left\{ 2 \int_B \left[[\int_B \phi y (G - G^*) dB]^2 + [\int_B G^* (\phi y - \phi y^*) dB]^2 \right] dB \right\}^{1/2}$$

$$\leqq \left\{ 2 \left[\int_B (\phi y)^2 \, dB \int_B \int_B (G - G^*)^2 \, dB \, dB \right. \right.$$

$$\left. \left. + \int_B (\phi y - \phi y^*)^2 \, dB \int_B \int_B G^{*2} \, dB \, dB \right] \right\}^{1/2}$$

$$< \{ 2 [C_{13} \, \epsilon_7 + C_{10} \, \epsilon_6] \}^{1/2}$$

according to (4.261,14), (4.261,30), (4.261,29), and (4.261,11), i.e., $\| \mathscr{B} - \mathscr{B}^* \|$ tends toward zero.
(4.260,vii):

$$\| \mathscr{C} - \mathscr{C}^* \| = \left\{ \int_B \left[(1-\nu) \int_{R_F} \left[\frac{\partial y^*_{x_1 x_2}}{\partial R} (G - G^*) + G \left(\frac{\partial y_{x_1 x_2}}{\partial R} - \frac{\partial y^*_{x_1 x_2}}{\partial R} \right) \right. \right. \right.$$

$$+ \frac{\partial G_{x_1 x_2}}{\partial R} (y^* - y) + G_{x_2 x_2} \left(\frac{\partial y^*}{\partial n} - \frac{\partial y}{\partial n} \right) + y^*_{x_2 x_2} \left(\frac{\partial G}{\partial n} - \frac{\partial G^*}{\partial n} \right)$$

$$\left. \left. \left. + \frac{\partial G}{\partial n} (y_{x_2 x_2} - y^*_{x_2 x_2}) \right] dR \right]^2 dB \right\}^{1/2}$$

$$\leqq \left\{ 6(1-\nu) \int_B \left[\int_{R_F} \left(\frac{\partial y^*_{x_1 x_2}}{\partial R} \right)^2 (G - G^*)^2 + G^2 \left(\frac{\partial y_{x_1 x_2}}{\partial R} - \frac{\partial y^*_{x_1 x_2}}{\partial R} \right)^2 \right. \right.$$

$$+ \left(\frac{\partial G_{x_1 x_2}}{\partial R} \right)^2 (y^* - y)^2 + G^2_{x_2 x_2} \left(\frac{\partial y^*}{\partial n} - \frac{\partial y}{\partial n} \right)^2 + y^{*2}_{x_2 x_2} \left(\frac{\partial G}{\partial n} - \frac{\partial G^*}{\partial n} \right)^2$$

$$+\left(\frac{\partial G}{\partial n}\right)^2 (y_{x_2x_2} - y^*_{x_2x_2})^2 \, dR \Bigg] \, dB \Bigg\}^{1/2}$$

$$< \Bigg\{ 6 \Bigg[\int_{R_F} \left(\frac{\partial y^*_{x_1x_2}}{\partial R}\right)^2 dR \iint_{B\,R_F} (G - G^*)^2 \, dB \, dR$$

$$+ \int_{R_F} \left(\frac{\partial y_{x_1x_2}}{\partial R} - \frac{\partial y^*_{x_1x_2}}{\partial R}\right)^2 dR \iint_{B\,R_F} G^2 \, dB \, dR$$

$$+ \int_{R_F} (y^* - y)^2 \, dR + \iint_{B\,R_F} \left(\frac{\partial G_{x_1x_2}}{\partial R}\right)^2 dB \, dR$$

$$+ \int_{R_F} \left(\frac{\partial y^*}{\partial n} - \frac{\partial y}{\partial n}\right)^2 dR \iint_{B\,R_F} G^2_{x_2x_2} \, dB \, dR$$

$$+ \int_{R_F} y^{*2}_{x_2x_2} \, dR \int_{R_F} \int_{B} \left(\frac{\partial G}{\partial n} - \frac{\partial G^*}{\partial n}\right)^2 dB \, dR$$

$$+ \int_{R_F} (y_{x_2x_2} - y^*_{x_2x_2})^2 \iint_{B\,R_F} \left(\frac{\partial G}{\partial n}\right)^2 dB \, dR \Bigg] \Bigg\}^{1/2}$$

$$< \{ 6 [C_6 \cdot \epsilon_8 + C_{18} \cdot \epsilon_4 + C_{22} \cdot \epsilon_2 + C_{19} \cdot \epsilon_5 + C_3 \cdot \epsilon_9 + C_{20} \cdot \epsilon_3] \}^{1/2}$$

according to (4.261,7), (4.261,31), (4.261,19), (4.261,27), (4.261,23), (4.261,25), (4.261,20), (4.261,28), (4.261,4), (4.261,32), (4.261,21), and (4.261,26), i.e., $\| \mathscr{C} - \mathscr{C}^* \|$ tends toward zero.

(4.260, viii):

$$\| \mathscr{D}^* \| = \Bigg\{ \int_B \Bigg[\int_B \Delta^2 y^* (G^* - G) \, dB + \int_{R_L} \Bigg[\Delta y^* \left(\frac{\partial G^*}{\partial n} - \frac{\partial G}{\partial n}\right)$$

$$+ y^* \left(\frac{\partial \Delta G^*}{\partial n} - \frac{\partial \Delta G}{\partial n}\right) \Bigg] dR + \int_{R_F} \Bigg[\frac{\partial \Delta y^*}{\partial n} (G - G^*)$$

$$+ \Delta y^* \left(\frac{\partial G^*}{\partial n} - \frac{\partial G}{\partial n}\right) \Bigg] dR \Bigg]^2 dB \Bigg\}^{1/2},$$

if one takes assumption (4.261,1) into account from the beginning on. Applying repeatedly the inequality of Schwarz as in the preceding cases, the result is

$$\|\mathscr{D}^*\| \leqq \Bigg\{5\Bigg[\int\limits_B (\Delta^2 y^*)^2\, dB \iint\limits_{BB} (G^*-G)^2\, dB\, dB$$

$$+ \int\limits_{R_L} (\Delta y^*)^2\, dR \iint\limits_{BR_L} \left(\frac{\partial G^*}{\partial n} - \frac{\partial G}{\partial n}\right)^2 dB\, dR$$

$$+ \int\limits_{R_L} y^{*2}\, dR \iint\limits_{BR_L} \left(\frac{\partial \Delta G^*}{\partial n} - \frac{\partial \Delta G}{\partial n}\right)^2 dB\, dR$$

$$+ \int\limits_{R_F} \left(\frac{\partial \Delta y^*}{\partial n}\right)^2 dR \iint\limits_{BR_F} (G-G^*)^2\, dB\, dR$$

$$+ \int\limits_{R_F} (\Delta y^*)^2\, dR \iint\limits_{BR_F} \left(\frac{\partial G^*}{\partial n} - \frac{\partial G}{\partial n}\right)^2 dB\, dR\Bigg]\Bigg\}^{1/2}$$

$$< \{5\,[C_9 \cdot \epsilon_7 + C_5 \cdot \epsilon_9 + C_1 \cdot \epsilon_{10} + C_8 \cdot \epsilon_8 + C_5 \cdot \epsilon_9]\}^{1/2}$$

according to (4.261,10), (4.261,30), (4.261,6), (4.261,32), (4.261,2), (4.261,33), (4.261,9), (4.261,31), (4.261,6), and (4.261,32), i.e., $\|\mathscr{D}^*\|$ tends toward zero.

One can conclude that the approximate eigenvalue λ^* calculated by means of the Quasi-Trefftz equations converges indeed toward the exact eigenvalue λ if the assumptions (4.261) are satisfied. This conclusion is correct as the assumptions (4.261) imply that the conditions of convergence (4.260) are satisfied.

5

The method of adjoint systems

5.1 Generalized principle of Hamilton

Dealing with nonconservative problems, the fundamental variational principle has to be applied in the form

$$\delta L = -\int_T \left[\int_B Q[w]\,\delta w\,dB + \int_R P[w]\,\delta w\,dR \right] dt, \tag{1.18}$$

i.e., $\delta L \neq 0$. Hence, it is not an extremum principle. An extremum principle would have been involved only if $Q \equiv 0$, $P \equiv 0$. However, these conditions were satisfied for conservative problems only.

By means of transformations mentioned in Section 1.2, the principle (1.18) can be brought into the form of the derived principle

$$\int_T \left[\int_B \left\{ \frac{d}{dt}\left(\frac{\partial \mathscr{T}}{\partial w_t}\right) + \frac{\partial U}{\partial w} \right\} \delta w\,dB \right.$$

$$= \left. \int_T \left[\int_B Q\,[w]\,\delta w\,dB - \int_R \{W(w, \delta w) - P[w]\,\delta w\}\,dR \right] dt. \tag{1.46}$$

From now on, only one-dimensional systems shall be considered. Then,

$$\mathscr{U} = \frac{1}{2} \sum_{k=0}^{2} h_k(x)\,[w^{(k)}(x,t)]^2. \tag{2.4}$$

According to (1.42) $\delta U/\delta w$ assumes the form

$$\frac{\delta U}{\delta w} = \sum_{k=0}^{2} (-1)^k\,[h_k w^{(k)}]^{(k)}. \tag{5.1}$$

Furthermore,

$$\frac{d}{dt}\left(\frac{\partial \mathscr{T}}{\partial w_t}\right) = Tw. \tag{2.2}$$

Using (2.2) and (5.1) in (1.46) yields

$$\int_T \left[\int_B \{Tw + \sum_{k=0}^{2} (-1)^k \, [h_k w^{(k)}]^{(k)}\} \delta w \, dB \right.$$

$$= \int_T \left[\int_B Q[w] \, \delta w \, dB - \int_R W(w, \delta w) - P[w] \, \delta w\} dR \right] dt. \tag{5.2}$$

By means of

$$w(x, t) = f(t) y(x) \tag{1.100}$$

and

$$Tf = -\lambda f, \tag{1.101}$$

(5.2) can be changed into

$$\int_B \{-\lambda y + \sum_{k=0}^{2} (-1)^k \, [h_k y^{(k)}]^{(k)}\} \delta y \, dB$$

$$= \int_B Q[y] \delta y \, dB - \int_R \{W(y, \delta y) - P[y] \delta y\} dR \tag{5.3}$$

after having suppressed in addition the factor $\int_T f^2 dt$. As one can see, the left hand side in (5.3) is formally selfadjoint. However, the selfadjointness of the relationship as a whole is obviously prevented by the nature of the nonconservative terms on the right hand side of (5.3). From this one may conclude that *nonconservative problems are not selfadjoint* and that the fundamental variational principle cannot be applied to them in its classical form as Hamilton's principle but rather in its form (1.18). That means, the fundamental variational principle is then no longer an extremum principle.

In the following it shall be shown that by introducing the space of *adjoint systems*, a *generalized principle of Hamilton* can be formulated according to which a certain functional, if not extremal, is at least stationary.

For example, a nonselfadjoint problem may be given as

$$Dw = 0, \tag{5.4}$$

$$(Gw)_R = 0, \tag{5.5}$$

where D is a linear differential operator and (5.5) are the boundary conditions. The adjoint system corresponding to (5.4), (5.5) reads

$$D^* v = 0, \tag{5.6}$$

$$(G^* v)_R = 0, \tag{5.7}$$

and is defined by the conditions

$$\int_{t_1}^{t_2} \int_0^l vDw\,dx\,dt = \int_{t_1}^{t_2} \int_0^l wD^* v\,dx\,dt, \tag{5.8}$$

$$\Big[\dot{w}v - w\dot{v}\Big]_{t_1}^{t_2} = 0. \tag{5.9}$$

If (5.4) and (5.5) are prescribed, the adjoint problem can be determined according to (5.8), (5.9) using integration by parts.

A well known example is *Beck's rod* shown in Figure 24, [50]. For this rod,

$$Dw = \mu\ddot{w} + \alpha w^{\mathrm{IV}} + Pw'' = 0, \tag{5.10}$$

$$(Gw)_R = w(0,t) = w'(0,t) = w''(l,t) = w'''(l,t) = 0. \tag{5.11}$$

Applying integration by parts to

$$\int_{t_1}^{t_2} \int_0^l v(\mu\ddot{w} + \alpha w^{\mathrm{IV}} + Pw'')\,dx\,dt,$$

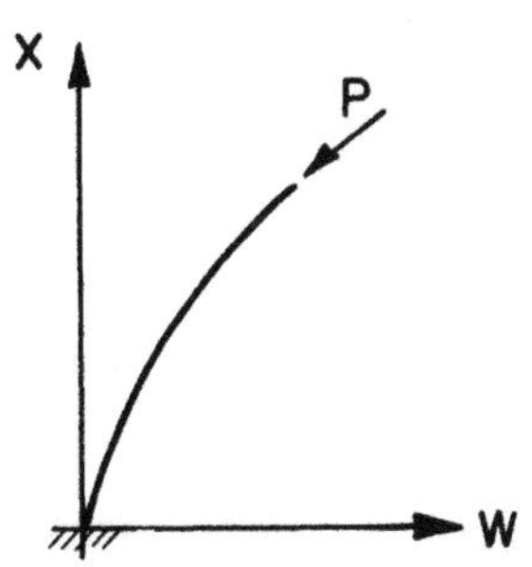

Figure 24. Beck's rod

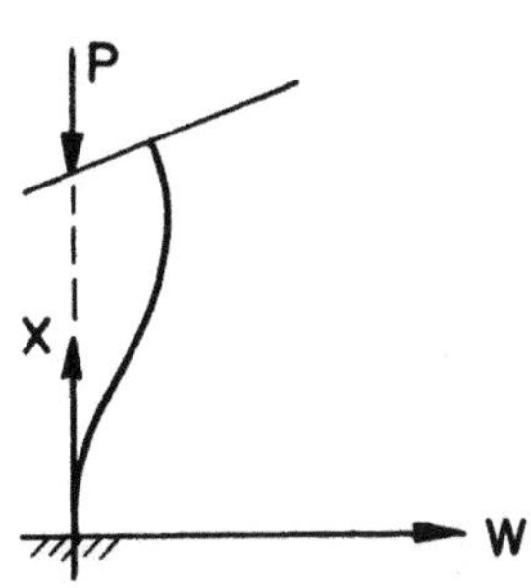

Figure 25. Reut's rod

and taking conditions (5.9) and (5.11) into account, one arrives at

$$\int_{t_1}^{t_2} \int_0^l v\,(\mu \ddot{w} + \alpha w^{\mathrm{IV}} + P w'')\,dx\,dt = \int_{t_1}^{t_2} \int_0^l w\,(\mu \ddot{v} + \alpha v^{\mathrm{IV}}\, P v'')\,dx\,dt \tag{5.12}$$

assuming in addition that

$$\begin{aligned}(G^* v)_R &= v(0,t) = v'(0,t) = \alpha v''(l,t) + P v(l,t)\\ &= \alpha v'''(l,t) + P v'(l,t) = 0.\end{aligned} \tag{5.13}$$

Comparing (5.12) with (5.8) yields

$$D^* v = \mu \ddot{v} + \alpha v^{\mathrm{IV}} + P v'' = 0, \tag{5.14}$$

i.e., the differential equation of the adjoint system, while (5.13) are its boundary conditions. In the sense of mechanics, the system adjoint to Beck's rod can be interpreted as *Reut's rod* as shown in Figure 25, [51].

Another example is *Leipholz' rod* in Figure 26, [52]. It is defined by

$$Dw = \mu \ddot{w} + \alpha w^{\mathrm{IV}} + [g(l-x)\,w']' + g w' = 0, \tag{5.15}$$

$$w(0,t) = w'(0,t) = w''(l,t) = w'''(l,t) = 0. \tag{5.16}$$

Taking condition (5.9) and boundary conditions (5.16) into account, the relationship

$$\begin{aligned}&\int_{t_1}^{t_2} \int_0^l v\,(\mu \ddot{w} + \alpha w^{\mathrm{IV}} + [g(l-x)\,w']' + g w')\,dx\,dt\\ &= \int_{t_1}^{t_2} \int_0^l w\,(\mu \ddot{v} + \alpha v^{\mathrm{IV}} + [g(l-x)\,v']' - g v')\,dx\,dt\end{aligned} \tag{5.17}$$

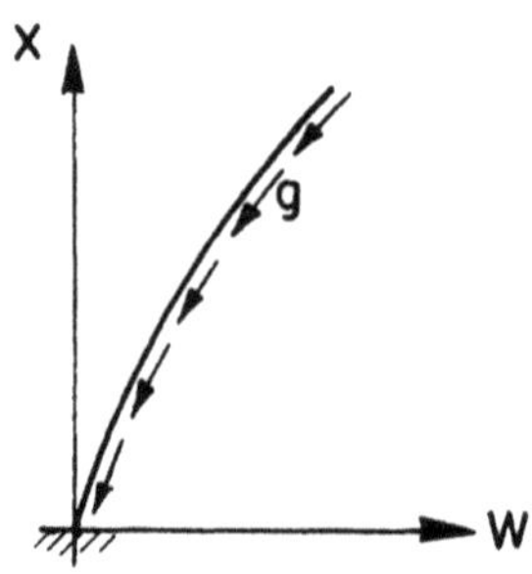

Figure 26. Leipholz' rod

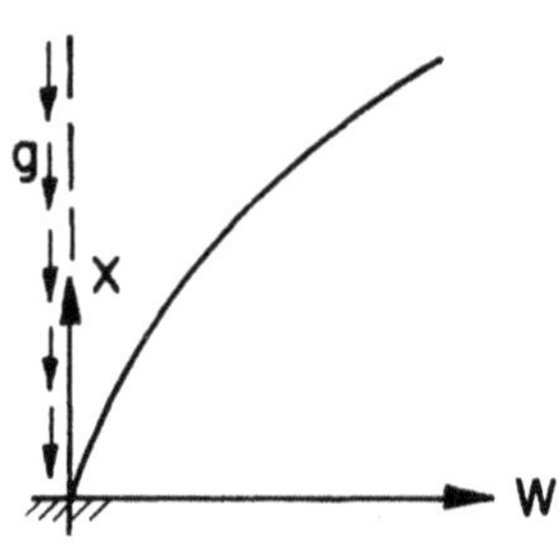

Figure 27. Barta's rod

is obtained by integrating by parts and setting

$$(G^*v)_R = v(0,t) = v'(0,t) = v''(l,t) = v'''(l,t) - \frac{1}{\alpha} gv(l,t) = 0. \tag{5.18}$$

A comparison of (5.17) with (5.8) shows that

$$D^*v = \mu\ddot{v} + \alpha v^{\mathrm{IV}} + [g(l-x)v']' - gv' = 0 \tag{5.19}$$

is the differential equation of the system adjoint to the system (5.15), (5.16). The relationship in (5.18) are the boundary conditions of the adjoint system, which can be interpreted in the sense of mechanics as *Barta's rod* shown in Figure 27.

For a system given in terms of $w(x, t)$, a *generalized Lagrange functional* can be formulated using the corresponding adjoint system in terms of $v(x, t)$. This is done in the following way: Firstly, a *bilinear kinetic energy*

$$T^* = \int_T\int_0^l \mathscr{T}^*\,dx\,dt = \int_T\int_0^l \mu\dot{w}\dot{v}\,dx\,dt \tag{5.20}$$

is introduced. Then, a *bilinear potential energy*

$$U^* = \int_T\int_0^l \mathscr{U}^*\,dx\,dt = \int_T\int_0^l \sum_{k=0}^{2} h_k(x)\,w^{(k)}v^{(k)}\,dx\,dt \tag{5.21}$$

is defined, and finally, the *bilinear work of the nonconservative forces*

$$A^* = \int_T\int_0^l \mathscr{A}^*\,dx\,dt = \int_T\int_0^l Q[w]v\,dx\,dt \tag{5.22}$$

is calculated. With these quantities, the bilinear, generalized Lagrange functional is constructed as

$$L^* = T^* - U^* + A^* = \int_T\int_0^l \mathscr{L}^*\,dx\,dt = \int_T\int_0^l (\mathscr{T}^* - \mathscr{U}^* + \mathscr{A}^*)\,dx\,dt, \tag{5.23}$$

where

$$\mathscr{L}^* = \mathscr{T}^* - \mathscr{U}^* + \mathscr{A}^* = \mu\dot{w}\dot{v} - \sum_{k=0}^{2} [h_k(x)w^{(k)}v^{(k)}] + Q[w]v \tag{5.24}$$

is the generalized Lagrange density.

For example, for Beck's rod, the Lagrange functional reads

$$L^* = \int_T \int_0^l \mathscr{L}^* \, dx\, dt = \int_T \int_0^l (\mu \dot{w} \dot{v} - \alpha w'' v'' + P w v' - \delta(x, l) P w' v) \, dx\, dt, \tag{5.23,1}$$

where $\delta(x, l)$ is Dirac's function. This expression has been obtained by generalizing in an obvious way the classical energies

$$T = \frac{1}{2} \int_T \int_0^l \mu \dot{w}^2 \, dx\, dt, \qquad U = \int_T \int_0^l (\alpha w''^2 - P w'^2) \, dx\, dt$$

in order to obtain the corresponding bilinear energies as

$$T^* = \int_T \int_0^l \mu \dot{w} \dot{v} \, dx\, dt, \qquad U^* = \int_T \int_0^l (\alpha w'' v'' - P w' v') \, dx\, dt. \qquad \begin{matrix}(5.20,1)\\(5.21,1)\end{matrix}$$

In addition, since $-Pw'(l)$ is the nonconservative force of the system, the corresponding bilinear work

$$A^* = - \int_T P w'(l) v(l) \, dt \equiv - \int_T \int_0^l \delta(x, l) P w' v \, dx\, dt \tag{5.22,1}$$

has to be considered. With all these terms, one can establish the functional (5.23,1).

For Leipholz' rod, one has to generalize

$$T = \frac{1}{2} \int_T \int_0^l \mu \dot{w}^2 \, dx\, dt, \qquad U = \frac{1}{2} \int_T \int_0^l (\alpha w''^2 - g(l - x) w'^2) \, dx\, dt$$

in order to obtain

$$T^* = \int_T \int_0^l \mu \dot{w} \dot{v} \, dx\, dt, \qquad U^* = \int_T \int_0^l (\alpha w'' v'' - g(l - x) w' v') \, dx\, dt. \qquad \begin{matrix}(5.20,2)\\(5.21,2)\end{matrix}$$

The nonconservative force of the system is $-gw'(x)$. Its bilinear work reads

$$A^* = - \int_T \int_0^l g w' v \, dx\, dt. \tag{5.22,2}$$

Using (5.20,2), (5.21,2), and (5.22,2) in (5.23) yields the generalized Lagrange functional

$$L^* = \int_T \int_0^l (\mu \dot{w} \dot{v} - \alpha w'' v'' + g(l-x) w' v' - g w' v)\, dx\, dt. \qquad (5.23,2)$$

For Hauger's rod [53] having the distributed load $g(l-x)^m$ and taking viscous damping into account,

$$L^* = \int_T \int_0^l \left(\mu \dot{w} \dot{v} - \alpha w'' v'' + \frac{g}{m+1} (l-x)^{m+1} w' v' - g(l-x)^m w' v'' - \beta \dot{w} v \right) dx\, dt, \qquad (5.23,3)$$

in order to give a more complicated example. In (5.23,3), β is the damping coefficient.

By means of the generalized Lagrange functional, a *generalized fundamental variational principle*

$$\delta L^* = \int_T \int_0^l \delta \mathscr{L}^* \, dx\, dt = 0 \qquad (5.25)$$

is defined, which corresponds completely to the classical principle of Hamilton, and which expresses the stationarity of L^*.

According to (5.23)

$$\delta L^* = \delta T^* - \delta U^* + \delta A^*. \qquad (5.26)$$

For δT^*,

$$\delta T^* = \int_0^l \int_T (\mu \dot{v} \delta \dot{w} + \mu \dot{w} \delta \dot{v})\, dx\, dt \qquad (5.27)$$

is obtained from (5.20). This expression can be changed into

$$\delta T^* = \int_0^l \{ [\mu (\dot{v} \delta w + \dot{w} v]_{t_1}^{t_2} - \mu \int_T (\ddot{w} \delta v + v \delta w)\, dt \} dx \qquad (5.28)$$

applying integration by parts. However,

$$[\delta w]_{t_1}^{t_2} = [\delta v]_{t_1}^{t_2} = 0. \qquad (5.29)$$

Therefore, (5.28) yields

$$\delta T^* = -\int_T \int_0^l \mu(\ddot{w}\,\delta v + \ddot{v}\,\delta w)\,dx\,dt. \tag{5.30}$$

By means of (5.29), (5.21), and (5.22), expression (5.26) becomes

$$\delta L^* = \int_T \Bigg[\int_0^l -\mu(\ddot{w}\,\delta v + \ddot{v}\,\delta w) + \delta\left(-\sum_{k=0}^{2} [h_k(x) w^{(k)} v^{(k)}] + Q[w]v\right) dx \Bigg] dt = 0, \tag{5.31}$$

taking in addition (5.25) into account. Now, assume that

$$\left.\begin{aligned} w(x,t) &= f(t)y(x), \\ v(x,t) &= f(t)z(x). \end{aligned}\right\} \tag{5.32}$$

Then, (5.31) yields

$$\int_T f^2\,dt \left[\left\{ \int_0^l -\mu \frac{\ddot{f}}{f}(y\,\delta z + z\,\delta y) + \delta\left(-\sum_{k=1}^{2} [h_k(x) y^{(k)} z^{(k)}] + Q[y]z\right) \right\} dx \right] = 0.$$

But, according to (1.101), (5.33)

$$\frac{\ddot{f}}{f} = \frac{Tf}{f} = \lambda.$$

Hence, using this fact, and suppressing $\int_T f^2\,dt$,

$$\delta \int_0^l \{\mu\lambda yz - \sum_{k=0}^{2} h_k(x) y^{(k)} z^{(k)} + Q[y]z\}\,dx = 0 \tag{5.34}$$

follows from (5.33). Introducing the functional

$$H^* = \int_0^l \{\mu\lambda yz - \sum_{k=0}^{2} h_k(x) y^{(k)} z^{(k)} + Q[y]z\}\,dx = 0, \tag{5.35}$$

$$H^* = \int_0^l \mathscr{H}^*\,dx,$$

respectively, having the density

$$\mathscr{H}^* = \mu\lambda yz - \sum_{k=0}^{2} h_k y^{(k)} z^{(k)} + Q[y]z, \tag{5.36}$$

the generalized fundamental variational principle can also be expressed in the form

$$\delta H^* = 0. \tag{5.37}$$

In analogy to the classical case, there exists also a *generalized derived variational principle:* According to (5.26), (5.27),

$$\delta L^* = \int_T \int_0^l [\mu\dot{v}\delta\dot{w} + \mu\dot{w}\delta\dot{v} - \delta\mathscr{U}^* + \delta\mathscr{A}^*]\,dx\,dt. \tag{5.38}$$

Moreover,

$$\int_T \int_0^l \mu\dot{v}\delta\dot{w}\,dx\,dt = \int_T \int_0^l \frac{\partial\mathscr{L}^*}{\partial\dot{w}}\delta\dot{w}\,dx\,dt \tag{5.39}$$

$$= \int_0^l \left[\frac{\partial\mathscr{L}^*}{\partial\dot{w}}\delta w\right]_{t_1}^{t_2} - \int_T \int_0^l \frac{d}{dt}\frac{\partial\mathscr{L}^*}{\partial\dot{w}}\delta w\,dx\,dt.$$

Using (5.29), one finds that

$$\int_T \int_0^l \mu\dot{v}\delta\dot{w}\,dx\,dt = -\int_T \int_0^l \frac{d}{dt}\frac{\partial\mathscr{L}^*}{\partial\dot{w}}\delta w\,dx\,dt. \tag{5.40}$$

Correspondingly,

$$\int_T \int_0^l \mu\dot{w}\delta\dot{v}\,dx\,dt = -\int_T \int_0^l \frac{d}{dt}\frac{\partial\mathscr{L}^*}{\partial\dot{v}}\delta v\,dx\,dt. \tag{5.41}$$

By means of transformations explained in Section 1.2, one arrives in addition at

$$\int_T \int_0^l \delta[\mathscr{U}^* - \mathscr{A}^*]\,dx\,dt = -\int_T \int_0^l \left(\frac{\delta L^*}{\delta w}\delta w + \frac{\delta L^*}{\delta v}\delta v\right)dx\,dt, \tag{5.42}$$

assuming that w *and* v *satisfy their respective boundary conditions*. The quantities $\delta L^*/\delta w$ and $\delta L^*/\delta v$ in (5.42) are variational derivatives, namely

$$\frac{\delta L^*}{\delta w} = \frac{\partial \mathscr{L}^*}{\partial w} - \frac{\partial}{\partial x}\frac{\partial \mathscr{L}^*}{\partial w'} + \frac{\partial^2}{\partial x^2}\frac{\partial \mathscr{L}^*}{\partial w''} \tag{5.43}$$

and

$$\frac{\delta L^*}{\delta v} = \frac{\partial \mathscr{L}^*}{\partial v} - \frac{\partial}{\partial x}\frac{\partial \mathscr{L}^*}{\partial v'} + \frac{\partial^2}{\partial x^2}\frac{\partial \mathscr{L}^*}{\partial v''}\,. \tag{5.44}$$

Using (5.40), (5.41), and (5.42) in (5.38) yields the generalized, derived variational principle in the form

$$\int_T\int_0^l \left\{\left[\frac{d}{dt}\frac{\partial \mathscr{L}^*}{\partial \dot{w}} - \frac{\delta L^*}{\delta w}\right]\delta w + \left[\frac{d}{dt}\frac{\partial \mathscr{L}^*}{\partial \dot{v}} - \frac{\delta L^*}{\delta v}\right]\delta v\right\} dx\,dt = 0, \tag{5.45}$$

if *all boundary conditions are satisfied.*
Due to the arbitrariness of the variations δw and δv, equation (5.45) yields the two *generalized Lagrange equations*

$$\frac{d}{dt}\frac{\partial \mathscr{L}^*}{\partial w} - \frac{\delta L^*}{\delta w} = 0 \tag{5.46}$$

and

$$\frac{d}{dt}\frac{\partial \mathscr{L}^*}{\partial \dot{v}} - \frac{\delta L^*}{\delta v} = 0, \tag{5.47}$$

which are the differential equations of the two adjoint systems.

Applying for example (5.47) to (5.23,3), the result is

$$\frac{d}{dt}\frac{\partial \mathscr{L}^*}{\partial \dot{v}} = \mu \ddot{w}$$

and, using (5.44),

$$\frac{\delta L^*}{\delta v} = -g(l-x)^m w' - \beta\dot{w} - \left[\frac{g}{m+1}(l-x)^{m+1} w'\right]' + [-w'']'',$$

which, according to (5.47) and after further rearranging terms, yields the differential equation

$$\mu\ddot{w} + \alpha w^{\mathrm{IV}} + \frac{g}{m+1}(l-x)^{m+1} w'' + \beta\dot{w} = 0 \tag{5.48}$$

of the damped Hauger rod. From (5.46), the differential equation of the adjoint system would be obtained correspondingly.

In an attempt to complete the analogy to the classical case, one could derive also *generalized canonical equations*. That has been done in [54].

5.2 Hybrid equations of Ritz

In order to solve the eigenvalue problem connected with nonselfadjoint problems, one may for example use the fundamental variational principle (5.37). Using (5.35), it reads

$$\delta \int_0^l \{\mu\lambda yz - \sum_{k=0}^{2} h_k y^{(k)} z^{(k)} + Q[y]z\} dx = 0. \tag{5.49}$$

Expanding y and z in terms of coordinate functions which satisfy all boundary conditions of the two adjoint systems correspondingly, one arrives at

$$y = \sum_i c_i \phi_i, \qquad z = \sum_i \gamma_i \psi_i. \tag{5.50}$$

Applying (5.50) to (5.49) and demanding as a consequence of (5.49) that

$$\frac{\partial H^*}{\partial \gamma_i} = 0, \qquad i = 1,2,3, \ldots, \tag{5.51}$$

the *hybrid equations of Ritz*, i.e.,

$$\sum_i c_i \int_0^l \{\mu\lambda \phi_i \psi_i - \sum_{k=0}^{2} h_k \phi_i^{(k)} \psi_j^{(k)} + Q[\phi_i] \psi_j\} dx = 0, \tag{5.52}$$

$$i, j = 1,2,3, \ldots,$$

are obtained. If one had satisfied the equivalent condition $\partial H^*/\partial c_i = 0$ instead of (5.51), one would have been led to a system of equations in place of (5.52), which would have had a matrix transposed to that of (5.52). The result *concerning the eigenvalues* would therefore have been the same as if one had used (5.52) for the calculations. Therefore, one may proceed with (5.52) for an *evaluation of the eigenvalues* without any loss of generality.

Observing that

$$h_1 = g h_1^* \tag{2.15}$$

and

$$Q[\phi_i] = -g\phi_i' \tag{5.53}$$

[see (1.64)], the system (5.52) can be changed into

$$\sum_i c_i\{\lambda A_{ij} - B_{ij} + gC_{ij}\} = 0, \tag{5.54}$$

where

$$\left.\begin{aligned} A_{ij} &= \mu \int_0^l \phi_i \psi_j \, dx, \\ B_{ij} &= \int_0^l [h_0 \phi_i \psi_j + h_2 \phi_i'' \psi_j''] \, dx, \\ C_{ij} &= -\int_0^l [h_1^* \phi_i' \psi_j' + \phi_i' \psi_j] \, dx. \end{aligned}\right\} \tag{5.55}$$

The system (5.54) is linear, algebraic, homogeneous. Therefore, it yields nontrivial coefficients c_i in the series expansion of y only if

$$\det(\lambda A_{ij} - B_{ij} + gC_{ij}) = 0 \tag{5.56}$$

holds true. Rearranging (5.56), one is led to the eigenvalue equation

$$F(\lambda, g) = 0, \tag{5.57}$$

whose graphical representation is an eigenvalue curve in the g, λ-plane. This curve has to be discussed in the usual way with regard to divergence and flutter loads:

$$F(0, g) = 0$$

yields divergence buckling loads,

$$\partial F/\partial\lambda = 0, \qquad F(\lambda, g) = 0$$

yields flutter loads.

As an example, choose Leipholz' rod. For this rod,

$$h_0 = 0, \quad h_2 = \alpha, \quad h_1^* = -(l-x).$$

Hence,

$$B_{ij} = \alpha \int_0^l \phi_i'' \psi_i'' \, dx,$$

$$C_{ij} = \int_0^l [(l-x)\phi_i' \psi_j' - \phi_i' \psi_j] \, dx.$$

Restricting oneself to $i, j = 1,2$ in (5.56), and introducing the notations

$$\left.\begin{aligned} G(g) = {} & (B_{11} - gC_{11})(B_{22} - gC_{22}) \\ & - (B_{21} - gC_{21})(B_{12} - gC_{12}), \\ H(g) = {} & - A_{11}(B_{22} - gC_{22}) - A_{22}(B_{11} - gC_{11}) \\ & + A_{12}(B_{21} - gC_{21}) + A_{21}(B_{12} - gC_{12}), \\ K(g) = {} & A_{11} A_{22} - A_{21} A_{12}, \end{aligned}\right\} \qquad (5.58)$$

equation (5.56) yields

$$\det(-\lambda A_{ij} + B_{ij} - gC_{ij}) = F(\lambda, g) = G(g) + \lambda H(g) + \lambda^2 K(g) = 0. \qquad (5.59)$$

As known [52], the Leipholz rod becomes unstable by flutter. For the calculation of the flutter load,

$$\frac{\partial F}{\partial \lambda} = H(g) + 2\lambda K(g) = 0 \qquad (5.60)$$

has to be used yielding the flutter frequency

$$\lambda_{fl} = -\frac{H(g)}{2K(g)}. \qquad (5.61)$$

Using (5.61) in (5.59) yields for the flutter load the equation

$$4G(g)K(g) - H^2(g) = 0. \qquad (5.62)$$

This equation shall be solved for

$$\left.\begin{aligned} \phi_1 &= x^2 - \frac{2}{3l} x^3 + \frac{1}{6l^2} x^4, \\ \phi_2 &= x^3 - \frac{1}{l} x^4 + \frac{3}{10l^2} x^5, \\ \psi_1 &= x^2 - \frac{5gl^3 + 24\alpha}{3l(gl^3 + 12\alpha)} x^3 + \frac{2gl^3 + 6\alpha}{3l^2(gl^3 + 12\alpha)} x^4, \end{aligned}\right\} \qquad (5.63)$$

$$\psi_2 = x^3 - \frac{7gl^3 + 120\alpha}{4l(gl^3 + 30\alpha)} x^4 + \frac{3gl^3 + 36\alpha}{4l^2 (gl^3 + 30\alpha)} x^5, \tag{5.63}$$

where the coordinate functions ϕ_i satisfy all conditions (5.16) and the ψ_i all conditions (5.18). The result is

$$\lambda_{fl} = 128.46 \frac{\alpha}{\mu l^4}$$

and

$$g_{fl} = 41.51 \frac{\alpha}{l^3}.$$

The value of the flutter load corresponds rather well to

$$g_{fl} = 40.7 \frac{\alpha}{l^3},$$

a value which has been calculated by means of the method of Galerkin [55].

5.3 Hybrid equations of Galerkin

In deriving the *hybrid equations of Galerkin*, one may start from the generalized derived variational principle (5.45). Since

$$\frac{\delta L^*}{\delta w} = -\frac{\delta U^*}{\delta w} + \frac{\delta A^*}{\delta w} \tag{5.64}$$

and

$$\frac{\delta L^*}{\delta v} = -\frac{\delta U^*}{\delta v} + \frac{\delta A^*}{\delta v}, \tag{5.65}$$

it may also be used in the form

$$\int_T \int_0^l \left\{ \left[\frac{d}{dt} \frac{\partial \mathscr{L}^*}{\partial \dot{w}} + \frac{\delta U^*}{\delta w} - \frac{\delta A^*}{\delta w} \right] \delta w + \left[\frac{d}{dt} \frac{\partial \mathscr{L}^*}{\partial \dot{v}} + \frac{\delta U^*}{\delta v} - \frac{\delta A^*}{\delta v} \right] \delta v \right\} dx\, dt. \tag{5.66}$$

Since

$$\mathscr{T}^* = \mu \dot{w} \dot{v}, \tag{5.67}$$

$$\mathscr{U}^* = \sum_{k=0}^{2} h_k w^{(k)} v^{(k)}, \tag{5.68}$$

and

$$\mathscr{A}^* = Q[w]\,v = -gw'v, \tag{5.69}$$

one obtains

$$\frac{d}{dt}\frac{\partial \mathscr{L}^*}{\partial \dot{w}} = \frac{d}{dt}\frac{\partial \mathscr{T}^*}{\partial \dot{w}} = \mu\ddot{v}, \quad \frac{d}{dt}\frac{\partial \mathscr{L}^*}{\partial \dot{v}} = \frac{d}{dt}\frac{\partial \mathscr{T}^*}{\partial \dot{v}} = \mu\ddot{w}, \tag{5.70}$$

$$\frac{\delta U^*}{\delta w} = \frac{\partial \mathscr{U}^*}{\partial w} - \frac{d}{dx}\frac{\partial \mathscr{U}^*}{\partial w'} + \frac{d^2}{dx^2}\frac{\partial \mathscr{U}^*}{\partial w''} = \sum_{k=0}^{2}(-1)^k[h_k v^{(k)}]^{(k)}, \tag{5.71}$$

$$\frac{\delta U^*}{\delta v} = \sum_{k=0}^{2}(-1)^k[h_k w^{(k)}]^{(k)}, \tag{5.72}$$

$$\frac{\delta A^*}{\delta w} = -\frac{d}{dx}\frac{\partial}{\partial w'}\,[Q[w]\,v] = gv', \tag{5.73}$$

$$\frac{\delta A^*}{\partial v} = \frac{\partial}{\partial v}\,[Q\,[w]v] = -gw'. \tag{5.74}$$

With these relationships, equation (5.66) becomes

$$\iint_{T\,0}^{\;\,l}\Bigg\{\left[\mu\ddot{v} + \sum_{k=0}^{2}(-1)^k[h_k v^{(k)}]^{(k)} - gv'\right]\delta w$$

$$+\left[\mu\ddot{w} + \sum_{k=0}^{2}(-1)^k[h_k w^{(k)}]^{(k)} + gw'\right]\delta v\Bigg\}dx\,dt = 0. \tag{5.75}$$

Due to the arbitrariness of the variations, this condition can be split up into the two independent conditions

$$\iint_{T\,0}^{\;\,l}\left[\mu\ddot{v} + \sum_{k=0}^{2}(-1)^k[h_k v^{(k)}]^{(k)} - gv'\right]\delta w\,dx\,dt = 0 \tag{5.76}$$

and

$$\iint_{T\,0}^{\;\,l}\left[\mu\ddot{w} + \sum_{k=0}^{2}(-1)^k[h_k w^{(k)}]^{(k)} + gw'\right]\delta v\,dx\,dt = 0. \tag{5.77}$$

Each of them yields a set of hybrid equations of Galerkin. For the evaluation of the eigenvalues it is sufficient to proceed with one of them. This may be (5.77). Using

$$\mu w + \sum_{k=0}^{2} (-1)^k [h_k w^{(k)}]^{(k)} + g w' = Dw, \tag{5.78}$$

it is possible to substitute (5.77) by

$$\int_T \int_0^l [Dw] \delta v \, dx \, dt = 0. \tag{5.79}$$

It is noteworthy that

$$Dw = 0 \tag{5.80}$$

is the differential equation of that system, which is given in terms of $w(x, t)$. Introducing again

$$w(x,t) = f(t) y(x), \quad v(x,t) = f(t) z(x), \tag{5.32}$$

equations (5.77) changes into

$$\int_T f^2 \, dt \int_0^l \left[\mu \frac{\ddot{f}}{f} y + \sum_{k=0}^{2} (-1)^k [h_k y^{(k)}]^{(k)} + g y' \right] \delta z \, dx = 0. \tag{5.81}$$

Using (1.101), i.e.,

$$c\ddot{f}/f = T\dot{f}/f = -\lambda,$$

and suppressing $\int_T f^2 \, dt$, equation (5.81) yields

$$\int_0^l \left[-\mu \lambda y + \sum_{k=0}^{2} (-1)^k [h_k y^{(k)}]^{(k)} + g y' \right] \delta z \, dx = 0 \tag{5.82}$$

as a foundation of the hybrid equations of Galerkin.

It may specifically be pointed out that (5.49) and (5.82) are completely equivalent. This can be seen in the following way: inserting $Q[y] = -gy'$ into (5.49) and after having performed the variation retaining the condition

$$\int_0^l \{ \mu \lambda y \delta z - \sum_{k=0}^{2} h_k y^{(k)} \delta z^{(k)} - g y' \delta z \} dx = 0, \tag{5.83}$$

which is an immediate consequence of (5.49) due to the arbitrariness and independence of the variations, equation (5.83) can be brought into the form of (5.82) integrating by parts.

The actual hybrid equations of Galerkin are obtained from (5.82) by means of the expansion (5.50) as

$$\sum_i c_i \left\{ \int_0^l \left[-\mu\lambda\phi_i + \sum_{k=0}^{2} (-1)^k [h_k \phi_i^{(k)}]^{(k)} + g\phi_i' \right] \psi_j dx \right\} = 0, \tag{5.84}$$

$i, j = 1,2,3, \ldots$.

For Beck's rod,

$h_0 = 0,\ h_1 = -P,\ h_2 = \alpha,\ g = \delta(x,l)P,\ \delta(x,l) = \text{Dirac function}.$

Hence, (5.84) yields

$$\sum_i c_i \{ \int_0^l [-\mu\lambda\phi_i + P\phi_i'' + \alpha\phi_i^{\mathrm{IV}} + P\delta(x,l)\phi_i'] \psi_j dx \} = 0. \tag{5.85}$$

By means of

$$I_{ij} = \mu \int_0^l \phi_i \psi_j dx,\quad B_{ij} = -\int_0^l [\phi_i'' + \delta(x,l)\phi_i'] \psi_j dx,\quad A_{ij} = \alpha \int_0^l \phi_i^{\mathrm{IV}} \psi_j dx, \tag{5.86}$$

relationship (5.85) changes into

$$\sum_i c_i \{A_{ij} - PB_{ij} - \lambda I_{ij}\} = 0. \tag{5.87}$$

This is a linear, algebraic, homogeneous system which yields nontrivial coefficients c_i only in the case that

$$\det(A_{ij} - PB_{ij} - \lambda I_{ij}) = 0 \tag{5.88}$$

is satisfied. Introducing the notations

$$\begin{aligned} \phi(P) &= (A_{11} - PB_{11})(A_{22} - PB_{22}) - (A_{21} - PB_{21})(A_{12} - PB_{12}), \\ \psi(P) &= -I_{11}(A_{22} - PB_{22}) - I_{22}(A_{11} - PB_{11}) + I_{12}(A_{21} - PB_{21}) \\ &\quad + I_{21}(A_{12} - PB_{12}), \\ K(P) &= I_{11}I_{22} - I_{21}I_{12}, \end{aligned} \tag{5.89}$$

and restricting oneself in (5.88) to $i, j = 1,2$, the further handling of (5.88) yields the eigenvalue curve

$$F(P,\lambda) = \phi(P) + \lambda\psi(P) + \lambda^2 K(P) = 0. \tag{5.90}$$

Condition

$$\frac{\partial F}{\partial \lambda} = \psi(P) + 2\lambda K(P) = 0$$

yields the flutter frequency

$$\lambda_{fl} = -\frac{\psi(P)}{2K(P)}, \tag{5.91}$$

which leads to

$$4K(P)\,\phi(P) - \psi^2(P) = 0 \tag{5.92}$$

if inserted into (5.90). Equation (5.92) may be used for the calculation of the flutter load P_{fl}.

Let $\alpha = 1$, $l = 1$ for the sake of simplicity, and let (5.92) be solved using the coordinate functions

$$\begin{aligned}
\phi_1 &= x^2 - \frac{2}{3}x^3 + \frac{1}{6}x^4,\\
\phi_2 &= x^3 - x^4 + \frac{3}{10}x^5,\\
\psi_1 &= x^2 - \frac{2(P^2+4P+24)}{P^2+6P+72}x^3 + \frac{P^2+12}{P^2+6P+72}x^4,\\
\psi_2 &= x^3 - \frac{2(P^2+12P+120)}{P^2+16P+240}x^4 + \frac{P^2+6P+72}{P^2+16P+240}x^5,
\end{aligned} \tag{5.93}$$

which satisfy the corresponding boundary conditions (5.11) and (5.13), then the result of the calculation of P_{fl} is

$$P_{fl} = 21.5$$

which agrees fairly well with the value $P_{fl} = 20.05$ given by Beck in [50].

5.4 Convergence considerations

As already mentioned, the hybrid equations of Ritz and the hybrid equations of Galerkin are equivalent with each other and can be mutually converted into each other by integrating the respective generating principles by parts. It is therefore sufficient to prove the convergence of one type of equations. The convergence of the other type of equations follows automatically. Convergence may be proved for the hybrid equations of Galerkin.

Let

$$\mu\lambda y + Dy = 0, \qquad G(y)_R = 0 \tag{5.94}$$

be the posed boundary value problem and

$$\mu\lambda z + \widetilde{D}z = 0, \qquad (\widetilde{G}z)_R = 0 \tag{5.95}$$

its adjoint problem. In order to generate suitable coordinate functions for the expansion of y and z, consider the "shifted" problem

$$D\phi = \tau\psi, \quad (G\phi)_R = 0, \tag{5.96}$$

$$\widetilde{D}(\psi) = \tau\phi, (\widetilde{G}\psi)_R = 0. \tag{5.97}$$

It may be assumed that there exists the inverse

$$D^{-1}\ldots = \int_0^l G(\rho, \eta)\ldots d\eta$$

of the operator D, where $G(\rho, \eta)$ is a Green function. Then, there exists also an inverse of $\widetilde{D}$, namely

$$\widetilde{D}^{-1}\ldots = \int_0^l G(\eta, \rho)\ldots d\rho.$$

As a consequence, equations (5.96) and (5.97) can be rewritten to yield

$$\phi(\rho) = \tau \int_0^l G(\rho, \eta)\,\psi(\eta)\,d\eta \tag{5.98}$$

and

$$\psi(\eta) = \tau \int_0^l G(\rho, \eta)\,\phi(\rho)\,d\rho. \tag{5.99}$$

Using (5.99) in (5.98) yields

$$\phi(\rho) = \tau^2 \int_0^l\int_0^l G(\xi, \eta)\,G(\rho, \eta)\,\phi(\rho)\,d\eta\,d\rho. \tag{5.100}$$

This is an integral equation having the *symmetric kernel*

$$K(\xi, \rho) = \int_0^l G(\xi, \eta)\, G(\rho, \eta)\, d\eta. \tag{5.101}$$

Therefore,

$$\phi(\xi) = \tau^2 \int_0^l K(\xi, \rho)\, \phi(\rho)\, d\rho \tag{5.102}$$

has a set of eigenfunctions ϕ_i which is complete in the space of the solutions of (5.94). The eigenvalues τ^2 of (5.102) are all positive, real and belong to a discrete spectrum.

Correspondingly, it can be shown, that there exists a set of eigenfunctions ψ_i, which is complete in the space of the solutions of (5.95). Moreover, there exists the bilinear expansion

$$G(\xi, \eta) \sim \sum_k \frac{\phi_k(\xi)\, \psi_k(\eta)}{\tau_k}, \tag{5.103}$$

which converges at least in the mean square [56].

The functions ϕ_i and ψ_i generated by the shifted problem shall now be used in the expansions

$$y = \sum_i \alpha_i \phi_i, \quad z = \sum_i \beta_i \psi_i. \tag{5.104}$$

These series converge, since y and z belong to the solution spaces for which the systems ϕ_i and ψ_i are complete.

Problem (5.94) can be written as

$$y(\xi) = -\mu \int_0^l G(\xi, \eta)\, y(\eta)\, d\eta \tag{5.105}$$

by means of the inverse of D. Using the expansions (5.103) and (5.104), equation (5.105) can be transformed into the approximative equation

$$\sum_k \alpha_k \phi_k(\xi) = -\mu\lambda \int_0^l \left(\sum_k \frac{\phi_k(\xi)\, \psi_k(\eta)}{\tau_k}\right)\left(\sum_i \alpha_i \phi_i(\eta)\right) d\eta. \tag{5.106}$$

This equation can be changed further in the following way: Comparing coefficients of ϕ_k in (5.106) yields

$$\alpha_k = -\mu\lambda \int_0^l \frac{\psi_k(\eta)}{\tau_k} \left(\sum_i \alpha_i \phi_i(\eta)\right) d\eta. \tag{5.107}$$

Multiplying both sides by τ_k results in

$$\alpha_k \tau_k = -\mu\lambda \int_0^l \psi_k(\eta) \left(\sum_i \alpha_i \phi_i(\eta)\right) d\eta. \tag{5.108}$$

However,

$$\alpha_k \tau_k \equiv \sum_i \alpha_i \tau_i \delta_{ik}, \tag{5.109}$$

and by definition,

$$\int_0^l \psi_i \psi_k \, d\eta = \delta_{ik}. \tag{5.110}$$

Hence, also

$$\alpha_k \tau_k = \sum_i \alpha_i \int_0^l D\phi_i \psi_k \, d\eta \tag{5.111}$$

by means of (5.109), (5.110), and (5.96). Using (5.111) in (5.108) yields

$$\sum_i \alpha_i \int_0^l D\phi_i \psi_k \, d\eta = -\mu\lambda \int_0^l \psi_k(\eta) \left(\sum_i \alpha_i \phi_i(\eta)\right) d\eta. \tag{5.112}$$

This can easily be transformed into

$$\sum_i \alpha_i \{ \int_0^l [\mu\lambda\phi_i + D\phi_i] \psi_k \, d\eta \} = 0, \tag{5.113}$$

which is the system of the hybrid equations of Galerkin corresponding to the problem (5.94). Thus, it has been shown that the hybrid equations of Galerkin are identical to the approximative equation (5.106) which can be written also in the form

$$y_n(\xi) = -\mu\lambda \int_0^l G_n(\xi, \eta) y_n(\eta)\, d\eta, \tag{5.114}$$

where

$$y_n(\xi) = \sum_{i=1}^{n} \alpha_i \phi_i, \tag{5.115}$$

$$G_n(\xi, \eta) = \sum_{k=1}^{n} \frac{\phi_k(\xi)\, \psi_k(\eta)}{\tau_k}. \tag{5.116}$$

The problem of the convergence of the hybrid equations of Galerkin is therefore identical to the problem of the convergence of the solution y_n to (5.114) toward the solution y to (5.105). In this sense the problem of convergence shall now be investigated.

Equation (5.105) can be represented as

$$y = -\mu\lambda T y \tag{5.117}$$

and equation (5.114) as

$$y_n = -\mu\lambda T_n y_n \tag{5.118}$$

by means of the operators

$$T \ldots = \int_0^l G(\xi, \eta) \ldots d\eta \tag{5.119}$$

and

$$T_n \ldots = \int_0^l G_n(\xi, \eta) \ldots d\eta. \tag{5.120}$$

In order to determine convergence, a theorem of S. G. Mikhlin [57] may be used. According to this theorem, the eigenvalues λ of (5.117) can be obtained from the eigenvalues of (5.118) by means of a limiting process for $n \to \infty$ if the operators T and T_n are completely continuous, and if $\| T - T_n \| \to 0$ for $n \to \infty$.

It shall now be shown that the conditions of Mikhlin's theorem are satisfied under certain conditions:

(i) The operator T is completely continuous if

$$\int_0^l \int_0^l G^2(\xi, \eta)\, d\xi\, d\eta < \infty. \tag{5.121}$$

The proof of this assertion can be found in [58].

(ii) The operator T_n is also completely continuous if (5.121) holds true. Proof: according to [58], T_n is completely continuous if

$$\int_0^l \int_0^l G_n^2(\xi,\eta)\,d\xi\,d\eta < \infty. \tag{5.122}$$

For the problem under consideration,

$$\int_0^l \left[G(\xi,\eta) - \sum_{k=1}^{n} \frac{\phi_k(\xi)\,\psi_k(\eta)}{\tau_k}\right]^2 d\xi = \int_0^l \left[G(\xi,\eta) - G_n(\xi,\eta)\right]^2 d\xi < \epsilon, \tag{5.123}$$

as pointed out in [56]. Expanding (5.123) yields

$$\int_0^l G^2(\xi,\eta)\,d\xi + \int_0^l G_n^2(\xi,\eta)\,d\xi < \epsilon + 2\int_0^l G(\xi,\eta)\,G_n(\xi,\eta)\,d\xi. \tag{5.124}$$

Squaring both sides in (5.124) and applying the inequality $(a+b)^2 < 2a^2 + 2b^2$ yields

$$\left[\int_0^l G^2(\xi,\eta)\,d\xi + \int_0^l G_n^2(\xi,\eta)\,d\xi\right]^2 < 2\epsilon^2 + 8\left[\int_0^l G(\xi,\eta)\,G_n(\xi,\eta)\,d\xi\right]^2. \tag{5.125}$$

Due to the inequality of Schwarz,

$$\left[\int_0^l G(\xi,\eta)\,G_n(\xi,\eta)\,d\xi\right]^2 < \int_0^l G^2(\xi,\eta)\,d\xi \cdot \int_0^l G_n^2(\xi,\eta)\,d\xi.$$

Therefore, (5.125) can be changed into

$$\left[\int_0^l G^2(\xi,\eta)\,d\xi + \int_0^l G_n^2(\xi,\eta)\,d\xi\right]^2 < 2\epsilon^2 + 8\int_0^l G^2(\xi,\eta)\,d\xi \cdot \int_0^l G_n^2(\xi,\eta)\,d\xi. \tag{5.126}$$

Rearranging further yields finally

$$\left[\int_0^l G^2(\xi,\eta)\,d\xi - \int_0^l G_n^2(\xi,\eta)\,d\xi\right]^2 < 2\epsilon^2 + 4\int_0^l G^2(\xi,\eta)\,d\xi \cdot \int_0^l G_n^2(\xi,\eta)\,d\xi. \tag{5.127}$$

Let

$$\int_0^l G_n^2(\xi, \eta)\, d\xi = X, \qquad \int_0^l G^2(\xi, \eta)\, d\xi = Y \tag{5.128}$$

for the sake of brevity. Then, (5.127) reads

$$\begin{aligned} &(Y - X)^2 < 2\epsilon^2 + 4YX, \\ &X^2 - 6YX + Y^2 - 2\epsilon^2 = g(X) < 0, \end{aligned} \tag{5.129}$$

respectively. Assume for example that $2\epsilon^2 = 0.1Y^2$, which is possible since ϵ is supposed to be an *arbitrary*, small, *positive* quantity. Then,

$$X_1 = 0.15Y > 0,$$

$$X_2 = 5.85Y > 0$$

are the two roots of $g(X)$. Since $g(0) \equiv Y^2 - 2\epsilon^2 > 0$, inequality (5.129) is satisfied by

$$0.15Y < X < 5.85Y.$$

Hence, also

$$0.15 \int_0^l Y\, d\eta < \int_0^l X\, d\eta < 5.85 \int_0^l Y\, d\eta. \tag{5.130}$$

Let

$$\int_0^l Y\, d\eta = \int_0^l \int_0^l G^2(\xi, \eta)\, d\xi\, d\eta = K,$$

K = positive, finite

in agreement with (5.121), then it follows from (5.128) and (5.130) that

$$0.15K < \int_0^l \int_0^l G_n^2(\xi, \eta)\, d\xi\, d\eta < 5.85K.$$

This inequality proves that (5.122) holds true if (5.121) holds true. Therefore, condition (5.121) implies in fact that T_n is completely continuous as has been asserted previously.

(iii) The condition $\| T-T_n \| \to 0$ for $n \to \infty$ is satisfied because of (5.123).
Proof:

$$\| Ty - T_n y \| = \left(\int_0^l \left[\int_0^l G(\xi,\eta) y(\xi) d\xi - \int_0^l G_n(\xi,\eta) y(\xi) d\xi \right]^2 d\eta \right)^{1/2}$$

$$= \left(\int_0^l \left[\int_0^l G(\xi,\eta) - G_n(\xi,\eta) \; y(\xi) \, d\xi \right]^2 d\eta \right)^{1/2}$$

$$\leqq \left(\int_0^l \left[\int_0^l G(\xi,\eta) - G_n(\xi,\eta) \; ^2 d\xi \cdot \int_0^l y^2(\xi) \, d\xi \right] d\eta \right)^{1/2}$$

$$= \left(\int_0^l \left[\int_0^l G(\xi,\eta) - G_n(\xi,\eta) \; ^2 d\xi \right] d\eta \right)^{1/2} \cdot \| y \| .$$

Consequently,

$$\| T-T_n \| \leqq \left(\int_0^l \left[\int_0^l \{G(\xi,\eta) - G_n(\xi,\eta)\}^2 d\xi \right] d\eta \right)^{1/2} = (l\epsilon)^{1/2}$$

according to (5.123). Since $\epsilon \to 0$ for $n \to \infty$, the previous assertion is correct.

By means of (i), (ii), and (iii) it has been shown that the theorem of Mikhlin is satisfied if (5.121) holds true. Hence, the hybrid equations of Galerkin yield a convergent approximation of the eigenvalue λ of (5.94) under this condition.

Should the posed problem be inhomogeneous, i.e., given by

$$Dy = f, \qquad (Gy)_R = 0, \tag{5.131}$$

then the adjoint problem reads

$$\widetilde{D}z = g, \qquad (\widetilde{G}z)_R = 0. \tag{5.132}$$

In (5.131), (5.132), functions f and g are supposed to be square integrable.

The hybrid equations of Galerkin

$$\sum_i \alpha_i \{ \int_0^l [D\phi_i - f] \, \psi_k d\eta \} = 0 \tag{5.133}$$

are equivalent to

$$y_n(\xi) = \int_0^l G_n(\xi, \eta) f(\eta)\, d\eta. \qquad (5.134)$$

The exact solution of (5.131) would be obtained from

$$y(\xi) = \int_0^l G(\xi, \eta) f(\eta)\, d\eta. \qquad (5.135)$$

The use of the hybrid equations of Galerkin is justified if it can be shown that

$$\int_0^l [y(\xi) - y_n(\xi)]^2\, d\xi \to 0 \text{ for } n \to \infty. \qquad (5.136)$$

Using (5.134) and (5.135),

$$\int_0^l [y(\xi) - y_n(\xi)]^2\, d\xi = \int_0^l \left[\int_0^l G(\xi, \eta) - G_n(\xi, \eta) f(\eta)\, d\eta \right]^2 d\xi \qquad (5.137)$$

is obtained. By means of the inequality of Schwarz one finds that

$$\int_0^l \left[\int_0^l \{G(\xi, \eta) - G_n(\xi, \eta)\} f(\eta) d\eta \right]^2 d\xi$$

$$\int_0^l \left[\int_0^l \{G(\xi, \eta) - G_n(\xi, \eta)\}^2\, d\eta \cdot \int_0^l f^2(\eta)\, d\eta \right] d\xi. \qquad (5.138)$$

Using this estimate for (5.137) it follows that

$$\int_0^l [y(\xi) - y_n(\xi)]^2\, d\xi$$

$$\leqq \int_0^l \left[\int_0^l \{G(\xi, \eta) - G_n(\xi, \eta)\}^2\, d\eta \cdot \int_0^l f^2(\eta)\, d\eta \right] d\xi. \qquad (5.139)$$

Because of (5.123), and because it was assumed that

$$\int_0^l f^2(\eta)\, d\eta = C < \infty,$$

$$\int_0^l [y(\xi)-y_n(\xi)]^2 \, d\xi \leqq Cl\epsilon$$

follows from (5.139). Since $\epsilon \to 0$ for $n \to \infty$, condition (5.136) is satisfied, and the convergence of the hybrid equations of Galerkin is established also for this case.

5.5 A stability theory of nonselfadjoint systems

Consider a linearly elastic body of volume Ω and surface S. Let $w(x_i, t)$, $i = 1,2,3$, be the prevailing deflection of the body from its position of equilibrium (for beams and plates for example the deflection normal to the axis, the middle surface, respectively), x_i are the spatial coordinates and t is the time. The equations of motion of the small vibrations of the body about its position of equilibrium be

$$\dot{w} = bp, \qquad \dot{p} = -Dw, \tag{5.140}$$

where $b = \mu^{-1}$, μ = mass density, p = generalized impuls, and D is a differential operator. The body be in a nonconservative state. Then, D is nonselfadjoint and corresponding to (5.140) there exists the adjoint system

$$\dot{v} = b\pi, \qquad \dot{\pi} = -\widetilde{D}v, \tag{5.141}$$

which is defined by

$$\int_\Omega (vDw - w\widetilde{D}v)\, d\Omega = 0. \tag{5.142}$$

Then, also,

$$\int_\Omega (vDp - p\widetilde{D}v)\, d\Omega = 0, \tag{5.143}$$

since $p = \mu\dot{w}$ is an admissible function which satisfies all the boundary conditions belonging to (5.140).

Let the bilinear functional

$$V = \int_\Omega (bvDw + \dot{v}\dot{w})\, d\Omega \tag{5.144}$$

be defined. It is invariant with respect to time. Hence,

$$\dot{V} = 0. \tag{5.145}$$

This fact can be shown as follows:

$$\dot{V} = \int_{\Omega} (b\dot{v}Dw + bvD\dot{w} + \ddot{v}\dot{w} + \dot{v}\ddot{w})\, d\Omega,$$

which due to (5.140) and (5.141) changes into

$$V = \int_{\Omega} (b^2 \pi Dw + b^2 vDp - b^2 p\widetilde{D}v - b^2 \pi Dw)\, d\Omega = 0,$$

taking also (5.143) into account.

This functional V is a *generalized Hamilton functional* and (5.140) are the corresponding canonical equations. Therefore,

$$Dw = \frac{\delta U}{\delta w} - Q[w] = \frac{\partial \mathscr{U}}{\partial w} - \frac{\partial}{\partial x_1}\frac{\partial \mathscr{U}}{\partial w_{x_1}} - \ldots + \frac{\partial^2}{\partial x_1^2}\frac{\partial \mathscr{U}}{\partial w_{x_1 x_1}} + \ldots,$$

where U is the potential energy, $\mathscr{U}$ its density, and $Q[w]$ the nonconservative force. Therefore,

$$\int_{\Omega} bvDw\, d\Omega = \int_{\Omega} b\frac{\delta U}{\delta w} v\, d\Omega - \int_{\Omega} bQ[w]v\, d\Omega. \tag{5.146}$$

If all boundary conditions are satisfied,

$$\int_{\Omega} b\frac{\delta U}{\delta w} v\, d\Omega = \int_{\Omega} b\left(\frac{\partial \mathscr{U}}{\partial w} v + \frac{\partial \mathscr{U}}{\partial w_{x_1}} v_{x_1} + \ldots + \frac{\partial \mathscr{U}}{\partial w_{x_1 x_1}} v_{x_1 x_1} + \ldots\right) d\Omega. \tag{5.147}$$

For a linear system, $\mathscr{U}$ is a quadratic function so that with suitable coefficients $C_0, C_1, \ldots, C_{11}, \ldots,$

$$\frac{\partial \mathscr{U}}{\partial w} = C_{ow}, \quad \frac{\partial \mathscr{U}}{\partial w_{x_1}} = C_1 w_{x_1}, \ldots, \quad \frac{\partial \mathscr{U}}{\partial w_{x_1 x_1}} = C_{11} w_{x_1 x_1}, \ldots, \text{etc.}$$

Using these expressions in (5.147) yields

$$\int_\Omega b\,\frac{\delta U}{\delta w}\,v\,d\Omega = \int_\Omega b\,(C_0\,wv + C_1\,w_{x_1}\,v_{x_1} + \ldots + C_{11}\,w_{x_1x_1}\,v_{x_1x_1} + \ldots)\,d\Omega. \tag{5.148}$$

Inserting (5.146) and (5.148) in (5.144) results in

$$V = \int_\Omega [b(C_0\,wv + C_1\,w_{x_1}v_{x_1} + \ldots + C_{11}\,w_{x_1x_1}\,v_{x_1x_1} + \ldots - Q[w]\,v) + \dot{w}\dot{v}\,]\,d\Omega. \tag{5.149}$$

This is indeed a generalized Hamilton functional which for conservative self-adjoint systems, for which $w = v, D = \widetilde{D}, Q\,[w] = 0$, changes into

$$V = \int_\Omega [b(C_0\,w^2 + C_1\,w_{x_1}^2 + \ldots + C_{11}\,w_{x_1x_1}^2 + \ldots) + \dot{w}^2\,]\,d\Omega = \frac{2}{\mu}\,H. \tag{5.150}$$

In (5.150), H is the *classical* Hamilton functional.

Systems (5.140) and (5.141) can be rewritten to yield

$$\ddot{w} + bDw = 0, \qquad (Gw)_S = 0, \tag{5.151}$$

$$\ddot{v} + b\widetilde{D}v = 0, \qquad (\widetilde{G}v)_S = 0, \tag{5.152}$$

including the respective boundary conditions. Let

$$\begin{aligned} w(x,t) &= e^{\omega t}y(x)\,. \\ v\,(x,t) &= e^{\sigma t}\,z\,(x), \end{aligned} \tag{5.153}$$

where y and z satisfy the boundary conditions in (5.151) and (5.152). Hence, from (5.151), (5.152) follows

$$\begin{aligned} &\omega^2\,y + bDy = 0, \\ &\sigma^2\,z + b\widetilde{D}z = 0. \end{aligned} \tag{5.154}$$

The eigenfunctions in (5.154) are known to be bi-orthonormal, i.e.,

$$\int_\Omega y_i z_j\,d\Omega = \delta_{ij}. \tag{5.155}$$

Because of the adjointness of D and $\widetilde{D}$,

$$\int_\Omega z_j D y_i d\Omega = \int_\Omega y_i D z_j\,d\Omega.$$

Consequently, using (5.154) and (5.155),

$$\omega_i^2 \int_\Omega z_j y_i \, d\Omega = - \int_\Omega b z_j D y_i \, d\Omega = - \int_\Omega b y_i \widetilde{D} z_i \, d\Omega = \sigma_j^2 \int_\Omega z_j y_i \, d\Omega,$$

$$(\omega_i^2 - \sigma_j^2) \int_\Omega z_j y_i \, d\Omega = 0 = (\omega_i^2 - \sigma_j^2)\, \delta_{ij}, \tag{5.156}$$

respectively. Since (5.156) is supposed to hold true for any value of i or j,

$$\omega_i = \sigma_i, \tag{5.157}$$

whereby (5.153) changes into

$$w(x,t) = e^{\omega t} y(x), \quad v(x,t) = e^{\omega t} z(x). \tag{5.158}$$

Now let the "shifted" problem

$$\begin{aligned} bD\phi &= \tau\psi, \qquad & (G\phi)_S &= 0, \\ b\widetilde{D}\psi &= \tau\phi, \qquad & (\widetilde{G}\psi)_S &= 0 \end{aligned} \tag{5.159}$$

be introduced corresponding to (5.154). As already mentioned in Section 5.4, the shifted problem yields the set of coordinate functions ϕ_i and ψ_i by means of which the convergent series expansions

$$w = e^{\omega t} \sum_i \alpha_i \phi_i, \tag{5.160}$$

and

$$v = e^{\omega t} \sum_i \beta_i \psi_i \tag{5.161}$$

are obtained.

Using (5.160) and (5.161) in (5.144), the result is

$$V = e^{2\omega t} \int_\Omega [(\sum_i \beta_i \psi_i)\, bD (\sum_j \alpha_j \phi_j) + \omega^2 (\sum_i \beta_i \psi_i)(\sum_j \alpha_j \phi_j)]\, d\Omega. \tag{5.162}$$

Using the notation

$$A_{kj} = \int_\Omega [\omega^2 \phi_j + bD\phi_j]\, \psi_k \, d\Omega, \tag{5.163}$$

equation (5.162) can be simplified to yield

$$V = e^{2\omega t} A_{kj} \beta_k \alpha_j, \tag{5.164}$$

applying in addition Einstein's summation convention. In the following, the factor $e^{2\omega t}$ shall be suppressed, and the calculation shall be continued with the bilinear form

$$\widetilde{V} = A_{kj}\beta_k \alpha_j. \tag{5.165}$$

Interpreting α_j, β_j as "coordinates" in a multiple-dimensional space, $\widetilde{V}$ can be developed about the origin of the coordinate system to yield

$$\widetilde{V} = \delta\widetilde{V} + \frac{1}{2}\delta^2\widetilde{V} + \ldots, \tag{5.166}$$

because $\widetilde{V}(0, 0) = 0$.

Let the *generalized principle of virtual work*

$$\delta\widetilde{V} = 0 \tag{5.167}$$

be introduced. Then, (5.166) changes into

$$\widetilde{V} = \frac{1}{2}\delta^2\widetilde{V} + \ldots. \tag{5.168}$$

In the sense of a *generalized energy criterion of stability* it may be demanded that $\widetilde{V}$ becomes indefinite, i.e., degenerates at the limit of stability. Because of (5.168) this requirement is equivalent with the condition that $\delta^2\widetilde{V}$ becomes indefinite (degenerates) at the limit of stability.

From (5.165) follows

$$\delta^2\widetilde{V} = 2A_{kj}\beta_k\alpha_j. \tag{5.169}$$

If the form on the right hand side of (5.169) is supposed to degenerate at the limit of stability, then, as well known in the theory of forms, this occurs when the discriminant of the form, i.e., the matrix of its coefficients becomes singular. Consequently,

$$\det A_{kj} = \det \frac{\partial^2\widetilde{V}}{\partial\alpha_i\partial\beta_j} = 0 \tag{5.170}$$

is the criterion indicating that the limit of stability has been reached.

The coefficients A_{kj} defined by (5.163) involve the frequency of the motion and the load parameter of the system. Therefore, a load-frequency relationship follows from (5.170). Let

$$\det A_{kj} = \det \frac{\partial^2 \widetilde{V}}{\partial \alpha_i \partial \beta_j} = F(g, \lambda), \quad \lambda = \omega^2, \quad g = \text{load}.$$

Then,

$$F(g, \lambda) = 0 \tag{5.171}$$

follows from (5.170). This is a hypersurface in the load-frequency space whose projection on the load planes yields the stability limits in these planes. The projections are obtained according to two conditions. The first reads

$$\det \left. \frac{\partial^2 \widetilde{V}}{\partial \alpha_i \partial \beta_j} \right|_{\lambda=0} = F(g, 0) = 0 \tag{5.172}$$

and yields the intersection of the F-hypersurface with the load plane. The second reads

$$\frac{d}{d\lambda} \left[\det \frac{\partial^2 \widetilde{V}}{\partial \alpha_i \partial \beta_j} \right] = \frac{\partial F}{\partial \lambda} = 0 \tag{5.173}$$

and is to be used together with (5.171). It yields the trace of that cylinder in the load plane which is tangential to the F-hypersurface and whose generators are parallel to the λ-axis.

The "curves" following from (5.172) and (5.173), (5.171) are the boundaries of the stability domains in the load planes. The first condition (5.172) yields those limits at which stability is lost by *divergence*. That can be seen as follows: Because of (5.172), λ, and therefore the frequency ω, is zero at this limit. Consequently, the mechanical system possesses at this limit a nontrivial *equilibrium* position. That is characteristic for divergence. The second condition (5.173) yields together with (5.171) the flutter boundary. The reason is that because of (5.173) and (5.171) the equations $F = 0$ and $F_\lambda = 0$ have a common root. That means $F = 0$ admits of a *double root*. Hence, at least one root λ of $F = 0$ is complex for load values beyond the boundary set by (5.173) and (5.171). Thus, ω became complex. Therefore, w would increase beyond any limit according to (5.158). This is flutter.

In order to justify the criteria (5.167) and (5.170), the following may be considered: For conservative systems, $\widetilde{V}$ is proportional to the potential energy up to one term which involves ω^2 and which is always positive. Therefore, (5.167) and (5.170) are essentially equivalent to the well known principles of the extremum and the positive definiteness of the potential energy which are used in the classical case to determine the equilibrium positions and the stability of conservative systems. Therefore, the theory developed here coincides completely with the classical theory of stability for conservative systems.

For nonconservative systems one has to argue in a different way: The system of hybrid equations of Galerkin corresponding to the posed problem

$$\omega^2 y + bDy = 0, \qquad (Gy)_S = 0$$

is

$$\sum_i \alpha_i \int_\Omega [\omega^2 \phi_i + bD\phi_i]\, \psi_k \, d\Omega = 0. \tag{5.174}$$

The derivation of this system of equations has been described in Section 5.4. Using (5.163) in (5.174) yields

$$\sum_i \alpha_i A_{ki} = 0, \tag{5.175}$$

which are the hybrid equations of Galerkin. On the other hand side, $\delta \widetilde{V} = 0$ is equivalent to

$$\frac{\partial \widetilde{V}}{\partial \beta_k} = A_{kj}\,\alpha_j = 0, \quad \text{(summation convention)}. \tag{5.176}$$

Hence, it becomes apparent that condition (5.167) yields the hybrid equations of Galerkin. Moreover, these equations admit of nontrivial solutions α_j only if according to (5.175)

$$\det A_{ki} = 0.$$

This condition is identical to (5.170). From all this it follows that the conditions (5.167), (5.170), on which the generalized theory of stability of nonselfadjoint systems is based, agree completely with the corresponding conditions of the hybrid method of Galerkin. Therefore, the generalized stability theory is well justified, because the convergence of the hybrid equations of Galerkin has already been proved in Section 5.4.

References

[1] GOLDSTEIN, H., *Classical Mechanics,* Addison-Wesley, Reading, Massachusetts/Palo Alto/London/Dallas/Atlanta, 1965, pp. 30–55
[2] PETTERSON, O., *Cirkulatorisk Instabilited Vid Tryckta Strävor Och Plattor,* Tekniska Högskolan j Lund, Institutionen för byggnadsstatik
[3] FÖPPL, A. and L., *Drang und Zwang,* Vol. 2, Oldenbourg, München/Berlin, 1944, p. 373
[4] SOBOLEV, S. L., *Partial Differential Equations of Mathematical Physics,* Pergamon Press, Oxford/London/Edinburgh/New York/Paris/Frankfurt, 1964, pp. 327–387
[5] COLLATZ, L., *Eigenwertaufgaben mit technischen Anwendungen,* Geest und Portig, Leipzig, 1963
[6] COLLATZ, L., *op. cit.* [5], pp. 144–149
[7] ZURMÜHL, R., *Matrizen,* Springer, Berlin/Göttingen/Heidelberg, 1958, p. 183
[8] KAMKE, E., *Differentialgleichungen, Lösungsmethoden und Lösungen,* Vol. 1, Geest und Portig, Leipzig, 1951, pp. 75*ff.*
[9] COURANT, R. and D. HILBERT, *Methoden der Mathematischen Physik,* Vol. 1, Springer, Berlin, 1931, p. 40
[10] COURANT, R. and D. HILBERT, *op. cit.* [9], pp. 39–47
[11] COURANT, R. and D. HILBERT, *op. cit.* [9], pp. 55–57
[12] COURANT, R. and D. HILBERT, *op. cit.* [9], p. 41
[13] COLLATZ, L., *op. cit.* [5], pp. 50*ff.*
[14] TRICOMI, F. G., *Integral Equations,* Interscience, New York, 1967, pp. 106–110
[15] COLLATZ, L., *Numerische Behandlung von Differentialgleichungen,* Springer, Berlin/Göttingen/Heidelberg, 1955, pp. 459–462
[16] COURANT, R. and D. HILBERT, *op. cit.* [9], pp. 45*ff.*
[17] TRICOMI, F. G., *op. cit.* [14], p. 125
[18] KANTOROWITCH, L. W. and G. P. AKILOW, *Funktionalanalysis in normierten Räumen,* Akademie-Verlag, Berlin, 1964, p. 145
[19] KANTOROWITCH, L. W. and W. I. KRYLOW, *Näherungsmethoden der höheren Analysis,* Deutscher Verlag der Wissenschaften, Berlin, 1956, pp. 41–43
[20] HAMEL, G., *Integralgleichungen,* Springer, Berlin/Göttingen/Heidelberg, 1949, p. 84
[21] COLLATZ, L., *op. cit.* [5], pp. 63–102
[22] LEIPHOLZ, H., *Stability Theory,* Academic Press, New York/London, 1970, pp. 95*ff.*
[23] LEIPHOLZ, H., *op. cit.* [22], pp. 205, 244
[24] LEIPHOLZ, H., (i), On Conservative Elastic Systems of the First and Second Kind, *Ing.-Arch.* 43 (1974), pp. 255–271; (ii), On Certain Nonconservative Elastic Systems Having Divergence Buckling Loads, *Mech. Res. Comm.* 1, 1974, pp. 245–249; (iii), A Generalization of Raleigh's Principle for Certain Nonconservative Eigenvalue Problems, *Development in Mechanics* 8, Proceedings of the 14th Midwestern Mechanics Conference, pp. 231–240
[25] COLLATZ, L., *op. cit.* [5], pp. 142–144
[26] KNOPP, K., *Theorie und Anwendung der unendlichen Reihen,* Springer, Berlin/Heidelberg, 1947, p. 353
[27] KANTOROWITCH, L. W. and W. I. KRYLOW, *op. cit.* [19], pp. 144–148

[28] COLLATZ, L., *op. cit.* [5], p. 71
[29] BISHOP, R. E. D. and D. C. JOHNSON, *The Mechanics of Vibrations*, University Press, Cambridge, 1960
[30] LEIPHOLZ, H., Über die Anwendung von Liapunows direkter Methode auf Stabilitätsprobleme kontinuierlicher, nichtkonservativer Systeme, *Ing.-Arch.* 39 (1970), pp. 357–368
[31] TRICOMI, F. G., *op. cit.* [19], pp. 95–98
[32] MAGNUS, K., *Schwingungen*, Teubner, Stuttgart, 1961, p. 59
[33] SOLODOWNIKOW, W. W., *Grundlagen der selbsttätigen Regelung*, Vol. 2, VEB Verlag Technik, Berlin/Oldenbourg, München, 1959, p. 912
[34] NIXDORFF, K., Ansätze bei der harmonischen Balance, *Regelungstechnik* 16 (1968), pp. 562–564
[35] NIXDORFF, K., Zur Konvergenz der harmonischen Balance im nichtregulären Fall, *Regelungstechnik* 16 (1968), pp. 506–509
[36] WULICH, B. S., *Einführung in die Funktionalanalysis*, Part 1, Teubner, Leipzig, 1961, p. 101
[37] COLLATZ, L., *Funktionalanalysis und Numerische Mathematik*, Springer, Berlin/Göttingen/Heidelberg, 1964, pp. 281*ff.*, 355
[38] LEIPHOLZ, H., Über die Konvergenz des Verfahrens der harmonischen Balance, *Regelungstechnik* 14 (1966), pp. 176–182
[39] HAMMERSTEIN, A., Nichtlineare Integralgleichungen nebst Anwendungen, *Acta Math.* 54 (1929), pp. 117–176
[40] LEIPHOLZ, H., Über das Verfahren von Grammel mit 'benachbarter' Greenscher Funktion und die entsprechende Verallgemeinerung des Verfahrens von Galerkin, *Ing.-Arch.* 37 (1968), pp. 260*ff.*
[41] GRAMMEL, R., Ein neues Verfahren zur Lösung technischer Eigenwertprobleme, *Ing.-Arch.* 10 (1939), p. 35
GRAMMEL, R., Über die Lösung technischer Eigenwertprobleme, *Forsch.-Hefte Stahlbau* 6 (1943), p. 36
BIEZENO, C. B. and R. GRAMMEL, *Technische Dynamik*, Vol. 1, Springer, Berlin/Göttingen/Heidelberg, 1953, pp. 177–181
[42] KOLMOGOROV, A. N. and S. V. FOMIN, *Measure, Lebesgue Integrals and Hilbert Space*, Academic Press, New York/London, 1961, pp. 113*ff.*
[43] KANTOROWITCH, L. W. and G. P. AKILOW, *op. cit.* [18], p. 145
[44] KNESER, A., *Die Integralgleichungen*, Vieweg und Sohn, Braunschweig, 1922, pp. 117–121
[45] COLLATZ, L., *op. cit.* [5], pp. 143*ff.*
[46] TREFFTZ, E., *Ein Gegenstück zum Ritzschen Verfahren*, Verh. Kongr. angew. Mech., Zürich, 1926, p. 131
[47] KNESCHKE, A., *Differentialgleichungen und Randwertprobleme*, Vol. III, Teubner, Leipzig, 1968, pp. 446–448, 452, 453
[48] LEIPHOLZ, H., *Einführung in die Elastizitätstheorie*, Braun, Karlsruhe, 1968, pp. 259*ff.*
[49] GOULD, S. H., *Variational Methods for Eigenvalue Problems*, University of Toronto Press, Toronto, 1957, p. 93
[50] BECK, M., Die Knicklast des einseitig eingespannten, tangential gedrückten Stabes, *ZAMP* 3 (1952), pp. 225, 476
[51] BARTA, J., *Acta Techn. Ac. Sc. Hungarical* 31 (1960), p. 241

[52] LEIPHOLZ, H., Die Knicklast des einseitig eingespannten Stabes mit gleichmäßig verteilter, tangentialer Längsbelastung, *ZAMP* 13 (1962), pp. 581–589
[53] HAUGER, W., Die Knicklasten elastischer Stäbe unter gleichmäßig verteilten und linear veränderlichen, tangentialen Druckkräften, *Ing.-Arch.* 35 (1966), pp. 221–229
[54] LEIPHOLZ, H., Über die Erweiterung des Hamiltonschen Prinzips auf lineare nichtkonservative Probleme, *Ing.-Arch.* 40 (1971), pp. 55–67
[55] LEIPHOLZ, H., *op. cit.* [22], pp. 239–241
[56] TRICOMI, F. G., *op. cit.* [14], p. 149
[57] MIKHLIN, S. G., *Variationsmethoden der Mathematischen Physik,* Akademie-Verlag, Berlin, 1962, p. 373
[58] LJUSTERNIK, L. A. and W. L. SOBOLEW, *Elemente der Funktionalanalysis,* Akademie-Verlag, Berlin, 1955, p. 147

Name and subject index

MIX
Papier aus verantwortungsvollen Quellen
Paper from responsible sources
FSC® C105338

If you have any concerns about our products,
you can contact us on
ProductSafety@springernature.com

In case Publisher is established outside the EU,
the EU authorized representative is:
Springer Nature Customer Service Center GmbH
Europaplatz 3, 69115 Heidelberg, Germany

Printed by Libri Plureos GmbH
in Hamburg, Germany